CODEMAN
BACK TO EARTH

By Scottland Little

Dedication

To my beloved grandmother, the architect of my dreams and my steadfast anchor, this book is lovingly dedicated to you. You were the first to see my autistic traits not as barriers but as windows to a vibrant imagination, shaping the way I perceive the universe and its infinite possibilities.

On every page of this journey through science fiction, I see reflections of the strength and encouragement you've provided. Your steadfast presence and understanding have been my sanctuary in the toughest times. This story, woven from the threads of otherworldly adventures, is imbued with the spirit of your love and the lessons of resilience you've taught me. To you, the first to know me and the last to judge, I dedicate this work. May this story carry a fraction of the guidance and light you've given me!

Table of Contents

CHAPTER 1
CHAMPIONS OF THE REALMS

In the vast omniverse, there were three realms: Techtopia, Netherheim, and EQ Haven. These realms were once united, but with the creation of Satan, the realms blamed each other for his evolution, leading them to drift apart, never to make contact again until now.

The celebrations of the reunion festivals could be heard throughout the galaxies. After Mathias, the prophesized hero of the omniverse, defeated Satan, the leaders of the realms made efforts to become allies once again. They started 'Project: RLMTRVL' to reconnect the realms and make them accessible by opening portals using complex engineering and teleportation rift spells.

After 12 months of research, trial, and error, the project was finally a success, and today was the grand opening of the first inter-realm portal. Bright fireworks colored the sky. Music could be heard from miles away, even in the Silent Forests of Netherheim, where no noise had ever been heard.

Codeman Mathias, was also happy. He was loving all the attention, but most of all, he was loving all the attention from Zeena, his new team member. She was the Elder's Granddaughter and Neveah's cousin. The Elder, who started it all, brought Mathias under his supervision and made him the true face of Codeman.

Everyone was happy, not knowing that the happiness would be short-lived. As the leaders of each realm gave speeches in their respective realms, announcing the grand opening, a fluctuation occurred, causing the portals to deactivate.

The Codeman, who was present in Techtopia to cut the crimson red ribbon, sensed that this was something more than a simple technical failure. Mathias clicked on his earpiece, "Aria, run the systems and check for any hack attempts or installed viruses."

Aria was the top student at EQ University and was brilliant with computers. She also practiced magic, allowing her to control shadows and blend with them. Aria investigated the logs on the portal back from the Crusaders HQ (Headquarters to an Organization that dealt with universal threats) but got blocked off.

"*I can't get through. Whoever caused the malfunction doesn't want us to find them,*" Aria informed Codeman.

Codeman sighed and thought for a moment before connecting to the realm leaders. It was frustrating to break the news to the leaders and ask them to postpone the grand opening to next week. The leaders were shocked to the core.

"Codeman, whoever is behind this had better be caught by then. We can't announce the grand opening and the incident without the culprit behind bars. It would panic the people," Empress Bofa, the leader of Netherheim, said worriedly. *"Speaking of the people, what about the crowd?"* asked Codeman. "We need to give them something else to distract them," he added.

"Listen, kid, how would you like to have your little team become the official protectors of our realms?" asked George Techsmith, the president of Techtopia, thoughtfully.

Codeman cringed inwardly at the word 'kid' but gave an affirmative nod and disconnected. Codeman, along with Techtopia's president, entered the stage. *"Citizens of Techtopia, we are proud to announce that before the opening of the inter-realm portal, we would like to introduce you to the person who made this all possible. The young man before you will also lead a team of superhuman individuals to protect our realms from tragedies like Satan. We are proud to announce the..."* said George, who now turned towards Codeman.

"The Champions of the Realm," Mathias whispered to George.

"THE CHAMPIONS OF THE REALMS!" yelled

George with enthusiasm.

George gestured to Mathias to take over the podium. Mathias took a deep breath before beginning his speech.

"Ladies and gentlemen, citizens across the realms, and all those who stand before me today, I, Codeman, and my team aim to stand as a symbol of unmatched power and unyielding justice. From the moment I was recruited by the Crusaders, it was clear that I was destined for something more than being a nomadic engineer. My every action, every breath, is for the betterment of the omniverse. I aim to be the embodiment of strength, the epitome of nobility, and the beacon of hope for all. It will be the Champions of the Realm's duty to protect and serve, to bring order to chaos, and to vanquish evil in all its forms. Our name shall be remembered throughout history as the ones who stood against wrongdoers who aimed to destroy the peace that is established once again, as the ones who dared to go against Satan. Heed my words and know that in the presence of Codeman and the Champions, justice will always prevail!"

After the festival came to a close, Codeman left the people for their celebration and hurriedly returned to HQ.

"How's the search going?" asked Mathias.

"No new leads yet," replied Alon.

Alon was the son of the previous general of the Crusaders. He was trained to take over his position ever since he was 12. Alon was a skilled pyromaniac (those who practice magic related to fire) and had the ability to make pyrokinetic weapons.

"Call the rest of the team to the meeting room," commanded Codeman.

Alon gave him a nod before heading off. Mathias arrived at the meeting to see his team, Alon, Aria, Amy, Neveah, Zeena, Robbie, and even Trophy, waiting for him. Only Lyra was nowhere in sight. Codeman noticed her absence but said nothing.

Once everyone was gathered in the meeting room, Codeman began, *"As you all know, the launch of Project TRVLRLM was disrupted by a cybernetic attack. We are given one week before the portal is booted back up to catch whoever is behind this."*

Mathias said before he was interrupted by Lyra barging in. "Sorry I'm late," sighed Lyra.

"Where were you?" snapped Codeman. *"Ever since you were freed from Satan, you never come to training, you are always late to meetings, and you never leave your room unless we have a mission. You used to be a hardworking, tough, and determined one, but it seems you only needed to avenge your parents, and now the current problems don't matter as much,"* he added.

Lyra apologized and took a seat at the table. *"Does anyone have any leads on who may be behind this? With them blocking us from being able to track them from the crash, we need more leads on a suspect,"* Codeman continued.

"Could be Blue-Glitch-" whispered Neveah.

"That's not possible. Alon put him in the dungeon after we were done with the interrogation," said Codeman.

"Don't put this on me. I thought Aria was going to do that," said Alon. *"Aria, can you cross-reference the portal's shutdown with Blue's virus?"* said Codeman Aria input the data into her tablet and processed it, *"It's a match,"* she said.

Codeman held his forehead disappointedly, asked Aria to trace his location, and called off the meeting.

Zeena followed Mathias to the hallway, *"Don't you think you were a bit hard on Lyra?"* she asked.

"What do you mean?" asked Mathias.

"You didn't have to call out Lyra in front of the whole team," Zeena replied.

"So, are you going to lecture me on my leadership? Don't forget that it's my leadership that defeated Satan in the first place," snapped Mathias as he marched away, leaving Zeena in the corridor.

Zeena entered the lobby, where Alon, Neveah, and Amy were in deep discussion. *"Is he always a jerk like that?"* asked Zeena.

"Who? Mathias?" asked Alon. Zeena gave an affirmative nod.

"Wouldn't blame him. It's hard being nice to a former criminal," Neveah said in a low voice, but Zeena heard and rolled her eyes.

"I'm going to go check on Aria," and with that, Zeena took off to Aria's lab.

"So, did you find Glitch?" asked Zeena, entering the lab.

"Whenever I get a location, the map updates, and the next second, he's somewhere else," replied Aria, widening her hands in exasperation.

"Do you think he corrupted our software?" asked Zeena. *"That's the current theory, I guess,"* Aria said dismissively.

Zeena felt as if she was unwanted; wherever she would go, no one seemed to want her there. *"Maybe Neveah was right,"* Zeena thought as she headed to bed.

That night, Zeena had a dream where she saw glimpses of what seemed to be her past. From all her middle school bullying days to all the crimes she had committed. Her practice of corrupted magic made her lose her humanity. She even saw her trial and rehabilitation process and how hard it was to overcome the corruption of corrupted magic. Zeena woke up in cold sweats. It wasn't the first time she'd had this dream. This has been happening every day since she came back to EQ Haven and joined the Crusaders. She left the bed and went to look out of the window. Neveah and Mathias were discussing something in the garden, their heads bent over a book. Zeena felt a pang in her heart. She looked at the time. It was after midnight. She shook her head, wondering why it was bothering her. She shrugged and came back to her bed to try to go back to sleep.

The morning came with a hustle. Aria had called a meeting as she had a potential lead on Blue-Glitch's location.

"What's up?" asked Alon as he entered the meeting room along with the other Champions.

"I have troubleshot all our software, and there is no diagnosis, and nothing is messing with our systems," said Aria. "So how is Glitch's location changing every time the map refreshes?" asked Zeena.

"I suspect that instead of infecting the portal with his virus to deactivate it, he used his virus to steal the data and inject himself with it. Thus, causing him to be able to travel across realms, dimensions, etc.," Aria replied.

"Then how are we supposed to take him down? Every time we can get close to him, he can simply escape," asked Codeman from his seat at the head of the table.

"*Why does he even want teleportation?*" asked Alon.

"*Since the festival, there have been some heists with no known suspect, no footage from surveillance, and no trace of the one who is responsible. My guess would be that this is what Glitch has been doing in his free time. If we can figure out a pattern in his heists, we can wait for him and catch him from there,*" replied Aria.

This was a plan that needed to be executed to retain the peace and harmony they had established with so many sacrifices and hardships. Mathias thought of Aunt Billie's last breath and clenched his fists in rage.

CHAPTER 2
BACK TO ACTION

There were only two days left until the relaunch of the inter-realm portals, and Codeman was growing more stressed.

It was astounding that Blue-Glitch was still not able to be tracked. Just a year ago, he was merely a rookie and merely a pawn in Satan's plan. Even though he had killed Lyra's parents during Satan's first attacks and captured Lyra in another battle, he still required a whole army of Deathbringers and barely managed to do so. The Champions even knew his weaknesses and strengths after researching him for so long. If only they had been more careful with him, he might have still been behind bars.

Codeman stood behind Aria as she coded her program to possibly predict his location. He was both surprised at how strong Blue-Glitch had gotten and disappointed in how they failed to contain him due to just a careless mistake.

Aria finally finished programming the software that could map out Blue-Glitch's heists.

"*Just need to input the data, process it, and… voilà!*" exclaimed Aria as her program executed, and her holographic screens showed three possible locations where Blue-Glitch could possibly turn up next.

"*So, where do we check first?*" asked Alon.

"*I think City Bank is most likely his target,*" suggested Neveah.

"*Or we could split up and stake out each location and report back any suspicions,*" said Zeena.

"*Great idea, split up we shall,*" said Mathias.

A shadow crossed Neveah's pretty face, and she gave Zeena an envious side-eye before Mathias broke the tension.

"*Robbie, Lyra, and Amy, you guys head to FantaTech Museum. Aria and Alon, you investigate City Bank, and Zeena and Neveah, you guys come with me to Magie-Appareil Jewelers,*" Mathias added.

The time was short. Codeman wanted to move fast. So, he directed the team to embark on this quest without losing a single minute.

While everyone was busy getting geared up, Zeena was in deep thought. Mathias' rage, mood swings, and hyper attitude were all confusing her. She had seen his composed and serene personality. In fact, he was a very humble person and a bit introverted when he was not leading the Champions as Codeman. She jerked out of thought when Codeman clapped his hands, propelling everyone to move fast. The team took their earpieces and split up into their three assigned groups.

Robbie, Lyra, and Amy entered and immediately mingled with the visitors at the museum. The building was an epitome of security due to the artifacts it treasured. The high ceilings met up forming a glass dome, a kind of screen where one could see the historic events associated with each artifact taking place to give in-depth knowledge to those who loved history.

On the other side of town, Alon and Aria stepped inside the City Bank. By now, everyone knew the Champions. Their identity evoked respect in every glance raised to them. Alon straightened spontaneously, cherishing the glory.

"I wonder how Blue-Glitch will corrupt City Bank's system. It's the most high-tech security system in the town," Alon wondered aloud.

"We know what he is capable of. I still think this site is a more potential target than the museum and the jewelers," Aria speculated, looking around for any unusual signs.

****

Zeena and Neveah were reveling in the glittering jewels.

"Ladies will be ladies," Mathias thought, rolling his eyes heavenward. Unlike the museum and the bank,

the jewelry showroom was built with ancient architecture, with murals of Egyptian, Greek, and Hindu bejeweled gods and goddesses on the wall. From outside, the showroom looked like a setting from an old movie. But, everyone knew it was just a ruse, a show to give a unique aura to the place. Each and every corner and case of the shop was secured with high-tech invisible radio waves.

With what seemed like hours of waiting, some action finally occurred at Magie-Appareil Jewelers. Mathias, Neveah, and Zeena noticed strange flickering in the store's light fixtures. The group decided to enter the main area to check if it was just an electrical issue or if there was something else afoot.

Upon entering the building, the whole store went quiet for a moment. Soft whispers broke the silence as the customers began talking to each other in hushed tones on the arrival of Codeman.

Zeena and Neveah thought it was a bit odd and creepy, but Mathias embraced the attention. However, he wasn't the only one who got distracted. Zeena hadn't been shopping in years since her imprisonment.

"Oh my God, that bracelet is so cute; this is totally something I would have stolen back in my days," said Zeena, glaring at a black bracelet made of crystal beads with a tint of purplish gleam to them. *"And apparently, it repels evil as well,"* she added.

"That would look really good on you," said Mathias.

"Really, right now? We are here to stop theft, not steal something ourselves," said Neveah, feeling another pang of jealousy at the tender tone Mathias used to praise Zeena. She was still struggling with her new feelings.

"I won't. I don't do that anymore, but it's really tempting not to," replied Zeena jokingly.

Their conversation, however, was short-lived as people came up to Codeman to get pictures, and soon Mathias lost track of what he was here for. Zeena and Neveah tried to pull Mathias out of the crowd, but they couldn't reach him with the hoard of fans crowding them, so they decided to split off.

"Jeez! These people are like a bunch of brainless zombies. They just crowd around anyone famous and completely ignore their privacy," said Zeena.

"Well, Mathias is a world-famous superhero and has saved these people countless times, not to mention he has a sort of charm with"

Neveah said before she was cut off by a loud noise followed by a blackout at the store.

"Guys, there's a blackout at the museum. I think this is where Glitch is pulling off the heist," radioed Lyra.

"That's odd. We have the same situation here," replied Aria through her radio.

"You think he has minions, or he might be duplicating," suggested Zeena.

"Well, I don't know what's the cause behind this, but I suggest we put a stop to whatever he has planned," said Alon before going radio silent.

Mathias caught up with the two after the blackout. *"What's going on?"* he asked.

"Kind of you to finally show interest," said Zeena.

Neveah and Zeena filled in Codeman on what had been going on, but in the corner of her eye, Zeena noticed Glitch before she lost him in the crowd. Mathias installed a scanner on his visor to scan and detect electrical signals to locate Blue-Glitch.

In the bank, Aria and Alon immediately went to the vault after the blackout to investigate. Alon could not see a thing due to the pitch darkness. He created a soft glow by making a fireball so that he could see what was going on in the vault.

Aria, the creature of darkness, had no such qualm. They saw Blue-Glitch looting bars of gold from the vault. Aria took advantage of the blackout as her abilities worked best in the darkness and summoned an army of shadow creatures

Shadows emerged from the depths of darkness, cornering Blue-Glitch against the vault wall, finally tackling him to the ground and pinning him. However, with his new abilities, Blue-Glitch teleported away, leaving Aria and Alon confused about what had happened. One moment Glitch was here, and the next he vanished.

"What was that? A holographic image?" Alon said, confused.

"I don't think so. It was very real. I think he was here and teleported to another location. He mastered the art sometime in the past months. That's the best answer I can deduce," Aria thoughtfully replied.

Luckily, Aria had placed a tracker which scanned his location, but to her surprise, the tracker indicated he was still where he was before he supposedly teleported. Aria removed the shadows, and Alon turned on the generator, where they saw the tracker still on the ground.

"Maybe the tracker didn't attach," suggested Alon, but Aria didn't believe so.

"The tracker was positively charged, so it would have attached to him, and with the strength of the charge, it would have a high range as well," Aria said thoughtfully.

"What's happening?" Alon tugged at his hair in confusion.

Back at the Jewelry store, Mathias tracked the signals to the security room where Blue-Glitch supposedly messed with the lighting. Mathias powered the lights back on, but there was no sign of Glitch, and no jewelry was robbed. He felt the same confusion that Alon and Aria were facing at the bank.

At the Museum, during the blackout, Robbie and Amy lost Lyra. Robbie went to turn the power back on, and as the museum lights once again filled the halls, Robbie and Amy saw Lyra and Blue-Glitch facing off each other.

Lyra took advantage of her abilities to control tech and turned on the teleportation blockers. However, Blue-Glitch was still teleporting. Lyra then used some turrets to fire at Blue-Glitch, but the bullets missed. Lyra thought it was because the tech was too old, but then she sensed another piece of technology. She closed her eyes and felt a more modern technology; she walked towards it and found a holographic projector. It was all an illusion, a diversion.

"Team, I think this was a set-up," radioed Lyra. Each of the groups found more holographic projectors.

"Teleportation, holographs, just how many tricks does this guy have under his sleeves?" Alon said with frustration in his voice.

"Well, if this was a distraction, then what was he distracting us from?" said Neveah.

"Maybe this," said Mathias as he received an alert on his phone of an infiltration at their HQ.

Mathias asked the Champions to return to HQ, where the team met back up. As they entered the HQ, they saw Blue-Glitch hovering out with a huge bulk of their latest weaponry and armor floating inside an electrical cage.

"Well, well, well, I guess a diversion can only last so long," chuckled and crackled Blue-Glitch with an evil grin on his face.

"You had a second chance, Glitch, you really wanna blow it like that?" Mathias asked sarcastically.

"Oh, please. This is my first chance to finally be my own person. To get out of that wretched Satan's shadow," said Blue-Glitch as he dropped the cage attached to his finger with an electric current string.

He opened it up and put his arm inside, absorbing all the data and tech through the same virus he used on the portals. This included Mathias's newest armor which supposedly had no weakness and was tested by the team. They had tested the armor by attacking it with each of their superpowers, and each was ineffective against the armor.

"Are you kidding me, there's no way we can stop him now," said Alon.

"There has to be some way. Maybe Lyra can use her powers to control Glitch," suggested Zeena.

"Mathias already made sure the armor was unable to be hacked or controlled by tech-controlling powers," said Lyra.

"Well, no one knows their creation's weakness better than the creator," said Zeena, looking at Mathias.

"Don't look at me, I promise you that armor is perfected, nothing can defeat it," Mathias the Codeman retorted.

"Clearly not. Blue-Glitch managed to infect it with his virus, so think of another weakness," replied Zeena.

"If you all are done standing around and chit-chatting. I want to test out my new abilities," Blue-Glitch came forth.

The team sighed before getting ready for battle. Alon started it off by throwing fireballs hotter than the hottest star, but it had no effect on Glitch. Neveah knew none of the attacks would have any use, so she used her telepathic abilities to tell the team a plan she made. She told Aria to code a virus and hide it in a device that Blue-Glitch would willingly absorb. The plan was their only chance, and the team agreed to distract Glitch while Aria did her thing. Codeman, however, was doubtful as he still believed his armor was unbeatable. Aria used her power to disappear into the shadows and sneak off to her lab. Lyra then used her Techmind ability to use the armor Blue-Glitch absorbed to fight him.

"Fight fire with fire is what I always say," said Alon. The fight went nowhere. To every attack, there was already a defense mechanism coded, and so it was

just back and forth. Aria came back with a huge handheld canon.

"Even with all the fail-safes, abilities, and protections, you still managed to forget to prepare the armor against-" Aria said before Zeena bumped into her, knocking the canon to Blue-Glitch.

Blue-Glitch gave a maniacal laugh and picked up the canon. *"You fools! It's a miracle you managed to take down Satan. You just aren't as good. Not that it matters now,"* Blue-Glitch absorbed the virus, letting it inside his system. The virus ate away all the data Glitch had stolen, reducing him to his original form.

Blue-Glitch was in obvious panic, buzzing and flickering with a speed like all the wires of a gadget had short-circuited. He was pushing every button he could lay his hands on, but not finding a way out of his defeat. Just when he was about to explode, Codeman came up with one of his own grievances.

"What a load of crap. No way a virus defeated my best armor," said Codeman. *"There were so many fail-safes and firewalls, not to forget the regeneration command to reboot the armor,"* he added.

"Regeneration command?" The word caught his ears, and Blue-Glitch instantly began looking for the solution that THE GREAT CODEMAN had so foolishly provided him in his arrogance.

As more and more data got eaten, Blue-Glitch looked for the command in his database, and with a moment's notice, he started gaining back all the data and ejected the virus.

"Thanks, Codeman. I thought you were smarter, or you just have a soft spot for me." He laughed at Codeman.

The team saw as Blue-Glitch teleported away, once again slipping out of their hands.

"Are you kidding me? We are tight on a deadline, and we let him escape," exclaimed Codeman, but no one responded. Instead, they all just gave him a look of disappointment over his arrogance before leaving the room.

Zeena was disappointed, but her mind was directing her to a very dreadful conclusion. Mathias the Codeman was not so arrogant, not so impulsive, and certainly not so foolish.

The news of the fake-out heists and Blue-Glitch's escape was all across the internet. Contrary to what the Champions had initially thought, there had been one theft.

"Bet it's a cool black bracelet in a certain jewelry store," said Neveah, eyeing Zeena, but they were surprised to find out it was at the FantaTech Museum.

"That's impossible," said Amy. *"Blue-Glitch wasn't even there, and after the blackout, the museum was*

evacuated, so there was no one else there except us," she added.

"Any details on what's stolen?" asked Aria.

"It's an artifact that was used by the wizard Zerlin to revive someone back from the dead. The artifact is known as Zerlin's Stone or the Revival Stone," Robbie explained.

"Who is Zerlin?" asked Alon.

"Zerlin was a famous wizard who was rumored to talk to the dead. He took advantage of his ability to make a profit and communicate with dead relatives or friends of his customers. The townsfolk soon discovered Zerlin was a fraud; he couldn't communicate with anyone he chose but only those who belonged in a plane called the Inbetween. Enraged by Zerlin's lies, the townsfolk took their revenge by killing Zerlin's family and exiling him. As revenge, Zerlin released the Undead to the town. But this couldn't bring his family back. Zerlin worked on the Revival Stone to bring his family back, but the ritual didn't go as planned, and even though his family returned, they were very sick. Their flesh would shed, and they would lose their minds until they became weak and frail. To make matters worse, they were unable to die by any means and lived to this day as mindless husks, stuck in this state for the rest of time."

So, who stole the stone if not Glitch? Everyone looked at each other. Another mystery, just what they needed to top up the problems.

CHAPTER 3
THE TRAP

The news about the theft of Zerlin's Stone soon reached the Elder. He immediately called the Champions, asking them to meet him in the meeting room.

The entire team gathered in the now very familiar meeting room, waiting for the Elder to arrive. An unseen tension surrounded the hovering table, swelling like a balloon ready to burst, which would disrupt the edgy silence of the room.

Codeman walked into the room to see everyone looking at him with irritated eyes and in complete silence. He ignored all the glaring eyes and took a seat. But his curiosity, or better to say, his arrogance, got the best of him.

"Why are you all looking at me like that? It's not my fault that the armor was built to perfection," he said.

Zeena's eyes started twitching, and she snapped, *"Mathias, can you quit being a jerk to the whole team? Why is it so hard to accept that you have just become arrogant and careless? You got easily distracted by the attention your 'Codeman' persona was getting at the jewelry store, and you let Blue-Glitch escape just because you didn't want your armor to look weak."*

Mathias realized his mistake and looked apologetic. He looked at his reflection on the cold glass table and then back at his team.

"I'm sorry," he said, but then he paused.

He again looked at himself and then at his team, clenching his fists. His sorry face now turned spiteful, and his eyes turned a shade of purple, very familiar to Zeena, and he continued, *"I'm sorry that saving your butts from Satan and leading this team isn't enough for you guys. Without me, there would be no peace, no reunion between realms. Hell, there might not have been any realms in the first place. If it weren't for me, you all would be dead. And this is the thanks I'm getting. And then you guys blame me for your inability to catch the bad guy."*

The team was caught off-guard. They felt themselves losing respect for Codeman. Mathias could see that, but he couldn't care less. However, Zeena finally knew what was going on with him. *"It's the corruption,"* Zeena thought to herself. The door to the meeting room opened, and the Elder entered, breaking the tension between the team.

"Greetings, children," the Elder said in his usual calm tone. He took a seat and began to address the issue at hand. *"I'm sure you kids have heard about the*

Book stolen from our library a year ago and also about the recently stolen Stone."

The team looked at one another before giving the Elder a nod.

"These items are involved with a form of magic. A form of magic that hasn't been seen or used for centuries due to its complexity and the consequences it brings. The objects are related to Necromancy," continued the Elder. *"Necromancy is the art of death. It has the ability to empower you to contact, manipulate, or even revive the dead,"* he added.

"Well, what's so wrong with revival?" asked Alon.

The Elder explained, *"Well, the thing with Necromancy is that it never fulfills its promises. The people that you contact or revive won't be how you knew them before death. And for every revival, there will be someone you have to sacrifice, to talk to someone you will have to offer your own soul. It's all about sacrifices."*

"So, what are we to do about it?" asked Mathias.

The Elder calmly took a sip of water before revealing,

"I believe that whoever is stealing all these artifacts is a follower of Satan and wants to revive Satan. I need you to investigate this thief before they do anything that destroys our newfound peace."

"Well, what about Blue-Glitch and the Opening of the InterRealm portal?" asked Lyra.

"Don't worry about that. I discussed it with the leaders, and we now have one more week. I want Alon, Aria, Lyra, Amy, and Robbie to go after Blue-Glitch," said the Elder.

"What about us?" asked Mathias.

"I want you, Neveah, and Zeena to focus solely on the investigation," the Elder said, concluding the meeting.

After the team left, Zeena stayed back with the Elder.

"Grandad, I think Mathias still-," Zeena said before she was interrupted by the Elder.

"Corrupted? Yes, I know. That's why I put you on the team with him and Neveah. You will not only be looking for the follower of Satan, but you also need to keep an eye on Mathias and find out how to fix him. Also, improve your relationship with Neveah. It's bad for the team," the Elder looked fondly at his granddaughter.

Zeena nodded and went to catch up with Mathias and Neveah. "I knew I recognized those eyes. Can't believe Mathias's corruption is returning, but then again it makes sense; he wasn't this arrogant when I first met him at the party when he defeated Satan. I should have put two and two together, but instead, I blamed *it on his fame. He was so sweet back then,"* Zeena muttered to herself as her cheeks went red.

She bumped into Mathias and Neveah, who were discussing their approach to get started on the investigation.

"So, what have you guys been planning?" Zeena asked.

"Well, I think it's obvious the first place to check is the most recent…The Museum," said Codeman, Mathias.

As the trio made their way to the museum, they couldn't help but get recognized. It didn't take long for them to get bombarded with a crowd, and soon even the paparazzi arrived to surround them.

Amidst the chaos, a woman yelled, *"Codeman, help."* Mathias immediately turned to see who it was but saw no one.

"Did you guys hear that?" asked Mathias, but Zeena and Neveah both shrugged in confusion.

Mathias looked around once more and finally shrugged it off, thinking it was just his imagination. The trio managed to sneak away from the crowd through an alleyway behind Italiano's Pizzeria.

"You attract quite the attention," said Zeena to Mathias.

"Well, he can't help the fact he's a well-known hero," replied Neveah, playfully giving Mathias a soft smile.

"Speaking of which, I'm sorry about what happened in the meeting room. It wasn't very heroic of me," said Mathias.

The two smiled at him and accepted his apology. Zeena wondered how Mathias's corruption effects worked. The whole time she thought he was consistently negative, but it rather seemed to be spontaneous bursts of negativity triggered by something. All she had to figure out was what, and then maybe she would be able to relieve him of his corruption. The trio finally made it into the Museum. There were laser cordons across the entrance, and detectives were on the scene.

Codeman approached the captain of the police team, *"Captain Joseph... May we have permission to investigate? We believe the theft is related to another more sinister crime we are investigating,"* said Codeman.

Be my guest. You are authorized by the government to assist in such crimes," said the captain. *"But Codeman, I think you should know that your team is on our suspect list as they were last on the scene. I thought I would just give you a heads-up,"* the captain informed Codeman with a smirk.

Codeman ignored the captain and entered the building. An officer guided them to the crime scene and explained what they had found so far. The officer also introduced the trio to the forensic scientist, Lincoln L. Thomas.

Lincoln Thomas was short and stout with receding hair. Surrounded by AI photographers and lab assistants, he looked like a very strange character but with a friendly smile and warm eyes.

"Ah, Codeman. A pleasure to meet you, my kids are huge fans."

"Hey Mr. Thomas, the pleasure is mine," replied Mathias, shaking his hand.

"Please, call me Link," Lincoln replied.

"So, what do we have here?" asked Zeena, eyeing a table full of evidence enclosed in special capsules.

"That's all the evidence we have gathered thus far. We managed to get a holographic projection of the scene along with some fingerprints and footprints. However, most of them are from the face-off with Glitch that happened last night. Our team is still looking for anything that can be helpful," replied Lincoln.

"Can we see the projection?" asked Neveah.

"Sure," said Lincoln, as he put on his gloves and pressed the button on the projector. Upon pressing the button, a blue light shone across the room before rendering a 4D replication of the scene. You could not

only see Lyra, Robbie, and Amy fighting the fake Blue-Glitch, but the details were so precise that you could even see a fly that was there on the scene. The details were so immersive that it made the training room look like nothing.

"*Crazy how technology is evolving after the peace*," said Codeman.

"*Well, ever since peace was established between realms and EQ Haven is back on the maps, there has been much more progress and collaboration. In fact, the Museum just bought these projectors and projection scanners from the most advanced security company in Techtopia*," Lincoln explained.

Hearing this really set in how much had changed since Satan was gone and made the trio even more determined to ensure he doesn't return. Lincoln went on to show the team how the stone was still there during the Glitch incident.

Then, suddenly, a steam-powered robot from the Steam-Punk era activated in the background with none of the team noticing as Blue-Glitch acted as the perfect distraction. The robot made its way to the stone and carefully lifted the case, then stole the stone and went back into position. After the fight ended and everyone had exited the building, the robot powered up again and escaped the Museum.

Could this be a hack?" asked Neveah.

"Initially, our team and I thought the same, but the supposed robot isn't actually a robot. The lack of sensors and ability to program makes it just a steam-powered humanoid machine. Thus, this cannot be hacked," replied Lincoln. *"However, what we do have is a tracker on the machine, and we can send you the data of the tracker. But the problem is the data has been corrupted and has been attempted to remove, so you will need to recover it first,"* he added.

"No problem!" said Codeman, as he sent the data over to Aria. *"Well, thanks for your time, Link,"* said Mathias, as the trio exited the Museum.

"Where to next?" asked Neveah.

"What about the library from where the book was stolen?" suggested Zeena.

"Doubt it, it's been too long since the theft from the library. If there was any evidence then someone would have noticed or found it already," said Neveah.

"I think it's still worth a shot," said Codeman, not knowing that Neveah was fuming inside for getting her say rejected once again. But she knew that Zeena had a valid point and she should've avoided the rebuttal just for the sake of opposing her idea.

The trio headed back to the alleyway to avoid any distractions.

"So, who do you think is our first suspect?" asked Mathias.

Well, I don't think we have any intel that points to a specific person," said Zeena.

"I think it's Glitch. He has the ability to hack anything and doesn't even have to be programmable. Plus, who else do we know who has the power to do such and was also a follower of Satan? Additionally, I don't think it's a coincidence the stone went missing when he started his crime spree, and he was also the distraction when it was stolen," said Neveah.

"True… But Glitch seemed pretty excited to be on his own again, and he is trying to get out of Satan's shadow," replied Zeena, ignoring Neveah's glare.

"Ladies, we're here," said Mathias. The trio entered the HQ library and went to the supposed crime scene.

"For an abandoned part of the ancient library… it's awfully clean," said Zeena.

"Well, I told you we won't find anything here," Neveah said triumphantly.

"Wait, check this out!" called Codeman, pointing towards the remnants of muddy footprints that were tried to be cleaned and weren't visible without Codeman's visors. *"Can you scan those?"* asked Zeena.

Codeman gave her a nod and scanned the prints with his visor. They were too old and smudged for the system to have a match, but it did pin down the date and time.

Wow, it got something as specific as date and time but not what shoe it is," said Zeena, rolling her eyes.

"Wait... that's because those aren't shoes. Those are Robbie's footprints," said Mathias.

"How is that possible, unless... do you think he's still corrupted by Satan from back when we rescued him at Techtopia?" said Neveah.

"I don't think that's it. Robbie had been recalibrated, and his whole system was redone after he broke during the final battle. I think we would have noticed any corruption. This is probably the same person who used that robot back at the Museum," said Zeena.

Mathias abruptly turned around.

"What happened?" asked Neveah.

"I thought I heard something," said Mathias. His eyes now looked over Neveah, and he rushed in that direction.

"Mathias, what in the world are you doing?" asked Neveah.

"How is this possible? They were just here," said Mathias.

"Who?" asked Zeena.

"The cloaked figure," said Codeman, still confused.

"*Mathias, I'm going to be really honest. You have lost it. There's nothing there,*" said Neveah, as she and Zeena walked into the room Mathias had dashed into.

As soon as they entered, the door shut behind them. Neveah tried opening it, but the door wouldn't budge... They were trapped.

****

CHAPTER 4
TROUBLED TRIO

Several hours had passed. The trio had no idea who had trapped them. Mathias punched the door in an attempt to break it down, but the door was made of Plathmatite, a strong metamorphic rock only used in ancient EQ Haven. Additionally, this library held books that belonged to the strongest engineers, wizards, and witches, meaning it had all sorts of protections against theft and intrusion.

Neveah, Zeena, and Codeman tried contacting the team for a rescue, but their radios were jammed by some sort of signal jammer.

"Ugh! It's been hours since we've been stuck here," groaned Neveah.

"I'm not enjoying this any more than you are," Codeman snapped.

"It's all your fault. You should have watched guard outside, and you are the reason we are here in the first place," said Neveah, pointing towards Zeena.

*"The hell do you mean? Mathias is the one who has been hallucinating all day. Come to think of it, you are

the only one who can tamper with minds. How do I know you aren't behind all this?" Zeena retaliated.

"Why would I do anything like that?" asked Neveah.

"Well, you have been a jerk to me ever since I joined the team. You fail to realize how people can change and only judge people based on their past. Do you realize how shallow that is? It isn't far-fetched that you would do all this just because you have some vendetta against me. All I have done is try to redeem myself to you, but you never gave me a chance," Zeena replied, her tone getting progressively more aggressive. Codeman, Mathias approached Zeena and circled her in his arms, calming her down with soothing words while glaring at Neveah.

Neveah turned around with tears of loneliness and dejection in her eyes. She had never felt lonelier in her entire life. She was also stuck, worried, and scared, but she had no one to support her. Above all, the man she was crushing on was glaring daggers at her for being harsh on another girl. She tried to compose herself, hiding her pain beneath an act of haughtiness.

The room fell into silence as Zeena started chanting spells to remain calm. She feared that if she lost her temper too much, the corruption would return.

While the trio was going through their dilemma, the remainder of the Champions were trying their best to take down Glitch. Tracking him wasn't very hard, as ever since he pulled off the heist at the Champions HQ, he felt invincible, boasting his power with complete transparency.

The Champions made their way to the Great EQ Garden, where Blue-Glitch was wreaking havoc.

Here's the plan. Aria and Amy, you clear out the civilians and make sure they are safe. Lyra and Robbie, you guys come with me to fight Glitch," said Alon as he peered over the park bench, analyzing Blue-Glitch.

"*Any strategies on how we do that?*" asked Robbie.

"*Plus, who made you in charge?*" asked Aria.

"*Since Codeman is gone, we need someone in charge, so I stepped up. Besides, we've got bigger fish to fry,*" said Alon.

As he finished his sentence, he heard a scream from behind him. "*And that's why we don't have time for a plan or a voting session to choose a substitute leader,*" he added as he clenched his wrist, making them engulf in flame.

Alon, Lyra, and Robbie charged toward Blue-Glitch, preoccupying him, while Amy turned down the power, causing all the lights to go out. With the park covered in darkness, Aria waved her fingers, chanting a spell to hide the bystanders in the shadows to evacuate them. However, due to the size of the crowd, the spell drained every ounce of her energy. While Aria was casting her spell, Robbie, Lyra, and Alon struggled to handle Blue-Glitch. Every time they came in his proximity, he teleported away. It had essentially become a game of cat and mouse, where the Champions would either chase Glitch or dodge one of his attacks. With each teleportation, Alon grew more frustrated until he had had enough. Alon blasted a huge fire in every direction, covering the whole park. Every tree, bush, and bench were on fire, lighting up the entire garden. The light emitted from the fire revealed the evacuating crowd. Glitch was stunned by the flames that had erupted all around him, but he quickly recovered and saw the evacuating people. Glitch paused, thinking of what better way to project fear in people's hearts and solidify himself as a true villain than to kill all the evacuating citizens. Glitch held his arm, making a fist, and charged up volts of electricity, ready to jolt towards the crowd. Everyone stopped what they were doing and realized what Glitch was about to do. Just when he was about to jolt towards the crowd, he stopped in his tracks.

A hooded figure appeared. The figure donned a very dark purple cloak with a hood, adorned with some green gems. Under the hood was complete darkness, except for two bright green eyes peering toward Glitch. The figure took out a device from under his cloak, a cylindrical tube with circuitry and wiring taped around it, indicating the person under the cloak was clearly an amateur. Glitch tried to move but seemed pinned by this new character. The figure flicked a loosely attached switch on the device, which pulled in all the electricity in the nearby area. And because Blue-Glitch was partially made of electrical energy, he was sucked in along with the electricity. The figure tucked the device into its belt and disappeared as mysteriously as it had appeared. Still in shock from what they had seen, the team regrouped.

"*What was that?*" asked Alon.

"*I don't know, but we got to go after it,*" replied Aria.

 Why? It seems that whoever it was, they have taken care of our jobs," said Lyra, panting as she ran into the group.

"*How do we know they weren't in cahoots? Can you really trust someone with no idea of their intentions?*" asked Aria. "*Besides that, where have you been?*" she added.

"*Ummm…err… I saw a kid crying, so I rescued him,*" Lyra replied in a soft voice. Lyra could tell that no one bought her story, particularly after noticing her tone and how much she stuttered. However, it didn't matter, as long as they didn't bring it up.

Back in the library, Mathias, Neveah, and Zeena had completely given up. They had tried every tool and spell, but the door was shut and sealed. It all seemed hopeless until Mathias heard a familiar barking. It was Trophy, Mathias's pet and childhood best friend.

"Finally, someone is here to rescue us," exclaimed Mathias with excitement in his voice.

"How is Trophy going to save us?" asked Neveah. But as soon as she said that, the door slowly opened, letting light shine into the dark, cold room. Mathias hugged Trophy and gave him pets.

"How do you manage to find me whenever I'm trapped in the most obscure places?" said Mathias, referring to the time Trophy had followed him into Satan's dungeon.

"Are we going to ignore how Trophy managed to open a completely sealed door without any spell or tool?" said Zeena.

Well, of course, he can. He's my pet, after all," Codeman answered, with the arrogance that was becoming a part of his personality now.

Zeena rolled her eyes and resolved to get to the bottom of this mystery. Sure, someone else had locked them in and later unlocked them, using Trophy as a distraction.

The trio, along with Trophy, headed back to HQ. Neveah kept trying to contact Aria for the tracker results but had no luck. The four of them made their way to Aria's lab to check the results for themselves and were shocked to find the tracker pointing towards their own HQ.

"Holy Sh-" and before Mathias could finish his sentence, a huge explosion was heard. They ran to where the noise came from and reunited with the rest of the Champions at Lyra's door, who had also rushed as soon as they heard the noise. Alon busted down the door to see Lyra in front of a huge summoning circle, with a Necromancy spell book hovering in the middle and Zerlin's gem in her hands.

"It was you all along," muttered Zeena."

"Of course, it had been me, ever since Satan promised a way to bring my parents back. I worked my butt off to pull this off, remaining unnoticed," Lyra spoke without any remorse.

"Lyra, this isn't you. You were our most determined and hardworking member. You were the one who always stood against wrong and made things right," Aria tried to reason with her.

You're wrong. I didn't care for justice. I wanted to avenge my parents. Don't you guys see it? All I did, have done, and will ever do, is for my parents," Lyra retorted.

"And you think they'll support you? Your parents were one of the first members of EQ Haven's secret force. You do know that to revive, you must kill. Will they ever be okay with knowing you sacrificed a life just for your own needs?" said Aria.

"It doesn't matter what they think. Besides, the sacrifice I have is better off dead," replied Lyra, as she pulled out the same cylindrical device that the hooded figure had.

"The hooded figure was you as well?" asked Alon.

"That's impossible. You were there with us when the figure appeared," said Amy.

"Wait, I also kept seeing a hooded figure throughout the investigation," said Mathias.

The dots were all connecting for Zeena. It appeared Lyra had been training to improve her psychic abilities. That is how she managed to manipulate minds, not only making everyone confused but also framing Neveah. It also explained how Lyra managed to control the steam-powered robot in the museum and how she bypassed Blue-Glitch's psychic block. Lyra pressed a button on the device, releasing Blue-Glitch in the middle of the room. Blue-Glitch, an advanced AI, was considered a living being by the spell. However, due to the armor and Blue-Glitch being only a fake copy of real intelligence, the process was taking longer. Alon tried to knock Blue-Glitch out of the circle, but his bionic prosthetic leg stopped working, making him stuck. Aria and Amy jumped at Lyra, but Lyra trapped everyone in their own minds while the process took place. Everyone fell to the ground except Neveah. Neveah's psychic abilities were canceling out those of Lyra. Neveah tried to retaliate by knocking

Lyra out with the same spell, but the two only just canceled each other out. Lyra became worried as Neveah was slowly gaining on her, and Lyra was starting to feel dizzy. So, in the moment, Lyra pulled out a syringe she had kept for emergencies such as this and injected it into Blue-Glitch before fainting. The syringe contained the Blue-Glitch anti-virus that they had developed back when the whole town was under the Blue-Glitch virus due to Satan's attack. This virus ate up Blue-Glitch from the inside. It was like cancer to him, except it was faster. And at a moment's notice, the ritual was complete. Blue-Glitch had perished, and a portal appeared. Everyone woke back up to see the portal of the in-between. Lyra's parents were summoned and were forced out into the real world. Lyra went up to hug her parents. Although her parents were glad to see their daughter all grown up, disappointment was evident on their faces for what she had done. As the team processed their defeat, Lyra celebrated her victory. The portal went unstable, and another set of feet walked out, leaving the team shocked as it was a face they hadn't seen in months and were hoping to never see again.

****

CHAPTER 5
REVIVAL

Everyone watched, stunned and wide-eyed, as the one who stepped out of the portal was none other than Satan. This time, however, he looked different. Satan's skin was rotting, and part of him was completely destroyed. It was surprising that he was even standing on his two feet. He looked around the room, seeing some familiar faces and some new ones, but one thing was common in all of them: shock and terror.

"Quite the reunion we have here," Satan said in his creepy, devilish voice as he slowly crept toward Codeman. Suddenly, Satan stopped in his tracks, rolled his eyes, and barged out the window, but no one followed. How could they? They were all in shock.

"What have you done?" Neveah said to Lyra.

"This is why necromancy is forbidden. It's unpredictable," added Zeena.

Then there was silence; palpable tension seeped into the room, leaving everyone speechless. The tension suddenly broke when the atmosphere crackled with a transmission message.

"*Apparently, there are skeleton-like creatures rising from the ground. They started appearing about the same time the portal opened,*" Robbie said, his database receiving information. "*These skeletons are hostile towards people. They also possess the ability to rearrange themselves and contort,*" he added, receiving more info.

"*Well, then we better head out,*" said Codeman.

"*But what about Satan?*" said Amy.

"*We'll deal with Satan after we ensure the civilians are safe,*" Codeman replied as he jumped out of the room from the same hole Satan had left. The team followed him out, leaving Lyra alone with her parents.

Lyra turned to face her parents once again, a tinge of regret reflecting on her face. Was bringing hell to earth just to bring her parents back really worth it? She looked at the hole in her room and then back at her parents, wondering if she should help her friends or stay.

"*They probably don't want to see me again anyway,*" Lyra thought out loud. Her mom comforted her by putting a strand of hair behind her ear and giving her a hug. At that moment, Lyra had everything she wanted, but she couldn't feel the happiness and contentment she had anticipated. And why were her parents not speaking?

"*Mum, Dad, are you alright?*" Lyra asked, feeling a slight pressure on the back of her shoulder, causing her to jump away. "Ouch." She checked where the pain was from, only to see scratches and grip marks on her shoulder. "*Jeez,*" Lyra shrieked.

Lyra turned back towards her parents to see their skin going abnormally pale, a helpless look of plea in their eyes—as if whatever was occurring within them was not in their control. They were slowly losing control but still trying to fight an urge. It was clear they were starting to fall under some influence, perhaps mind control or even simple instincts. Could this be what Robbie had mentioned about the Zerlin family? Paralyzed by the revelation, Lyra couldn't move or believe that all her efforts, risks, and sacrifices had been for nothing. Every second that passed, her parents lost more control. They banged their bodies against the wall in an attempt to stop themselves from harming their daughter. In that moment, Lyra snapped back to consciousness and rolled over, seeing her parents crash into the wall.

"They might be hostile, but they're still my parents. I need to get them somewhere safe where they can't hurt anyone," she mumbled to herself. Lyra ran out of the room with her parents trailing behind. She led them into the cell rooms, directing her parents into the cells as well. As her parents approached her, Lyra accepted fate and used her powers of controlling technology to activate the laser gate, entrapping herself with her parents.

Meanwhile, the Champions reached the scene and were mortified by what was happening. Skeletons were all over town, relentlessly tormenting anything that moved. Assuming the skeletons would be easy to deal with, the Champions decided to fight them off but were instantly proven wrong. Mathias punched a skeleton with the full force of his hydraulic titanium metal gauntlet, shattering it. However, they soon found that the skeletons could rebuild themselves. Codeman's actions drew attention towards the team, and a horde of skeletons began approaching.

As the skeletons marched towards the Champions, ready to face whatever they threw at them, they started disassembling their bones, rearranging them, and merging with others to become threatening gladiators. Some turned into large, bulky brutes, while some turned into centaurs, all wielding medieval weapons. These skeletal creatures were unlike anything anyone had ever seen. Despite being well-versed in dealing with all types of AI-based gadgets and advanced weapons, the Champions were overwhelmed by armories they had only seen as relics in ancient books.

As the team nervously started backing away, Alon asked, *"What's the plan, Matty?"*

"The skeletons outnumber the whole town, and they clearly are full of surprises. We can't save them all," muttered Codeman in a shaky, anxious voice.

However, Codeman, too, suppressed his worries, recollected his thoughts, and cooked up a plan: *"Aria and Robbie, get the inter-realm portal working. Alon and Neveah, make sure there are no skeletons around the area. Amy and Zeena, you and I are going to gather civilians. Take anyone you can find while the portal is being fixed,"* instructed Codeman.

"Wait, so we are just going to run? We're going to leave our homes and not even put up a fight?" asked Alon.

"Alon, look around you; this place is not safe. It's covered with creatures who are strong, unpredictable, and hell-bent on hurting people. And more of these are rising every minute. We can't fight this. The next best thing is to run, to ensure the civilians are safe," Codeman explained, and Zeena was thankful that Codeman was his usual self at this crucial moment.

Alon gave an affirmative nod, and the team split up into their respective roles. Aria pulled out a cube from her pocket, which unfolded into a hoverboard, and she and Robbie rushed to the portal.

"Robbie, download the portal schematics," Aria said.

As they reached the portal, Aria immediately got to work. She plugged Robbie into the portal's mainframe and began attempting to fix the portal. Alon and Neveah circled and scouted the perimeter to ensure the area was skeleton-free. Luckily, thus far, nothing was off. Amy, Mathias, and Zeena split up to cover as much ground as possible.

Amy heard screaming coming from inside a hospital and quickly rushed in. Exploring the seemingly evacuated building, she advanced towards the screams. Approaching the ER room, she saw a group of skeletons surrounding a door. The skeletons kept banging and bashing until one noticed Amy. As that one skeleton approached her, the others quickly followed. Amy shot the approaching skeletons, causing them to tumble like a Jenga tower, but that would only slow them down as they quickly rebuilt themselves. Amy ran towards the psychiatric ward of the hospital as the skeletons chased her. Soon, she was backed into a corner, the strange skeletons getting closer. She swung her fists in desperation and threw them into the padded cells of the psychiatric ward, locking them in. Wiping the sweat off her forehead, she went back to the ER room.

"It's safe now," she yelled to the survivors inside. The door opened, revealing a small family. There was a woman in her late 40s, two small children, and a boy who looked 20. The woman was injured, with blood gushing from her leg wound, splattering all over the place.

"Please help her. My mother...she was attacked by one of the monsters and can't walk," the boy rushed toward Amy for help. Amy looked at the family with

sympathetic eyes and tried to calm them. Settling their nerves with soothing words, she inspected the wound, asked for some medical supplies, treated the wound, disinfected it with a laser beam, and bandaged it. The family walked out of the room, with Amy grabbing a wheelchair for the woman.

Meanwhile, Aria had almost fixed the portal, but unfortunately, there was a part missing. Punching the ground in frustration, her mind raced for a solution. Suddenly, an idea struck her, and she quickly got up and radioed Alon to fetch the part from her lab.

On the other end of town, Mathias roamed the streets and managed to gather a good number of survivors. It seemed the skeletons had not reached this area of town yet. But that didn't last long as a horde of them approached from behind. Mathias, Codeman, blasted towards the skeletons with his hand-cannons, but that would only slow them down. He asked the survivors to run, as there was no use fighting such a huge army.

As the survivors ran, Mathias looked around, saddened to his core. "*Why did you have to do that, Lyra?*" he mumbled to himself and turned around to drive everyone towards the inter-realm portal.

Zeena, however, still had no luck finding any survivors. Everywhere she went, the skeletons had already been there. Finally, she heard a yell for help from the prison. Entering the building full of skeletons surrounding the prisoners' cells, Zeena rang the prison bell, alerting the skeletons to her presence. Now cornered, with perhaps hundreds of skeletons advancing towards her, Zeena grabbed a police hand cannon and started firing, but as usual, it was no use. With no other option, she clenched her fists and muttered a spell under her breath; her eyes turned purple, and soon enough, the skeletons started to turn black; their bones withered and turned into dust. Zeena had used an ounce of corrupted magic to destroy the skeletons. Now, with the horde gone, she rummaged through the drawers for any keycard but found none. She made her way to the retina scanner, used corrupted magic to morph her eye into one of the officers', and unlocked the prison. The laser beams serving as the prison bars for the prisoners vanished, and she released all the prisoners.

"Follow me if you want to live, but no funny business, or else you go straight to the skeletons. Am I making myself clear?" Zeena asked. The prisoners nodded.

"But why? All the officers evacuated. They left us here just because we are criminals," one of the prisoners spoke up.

"Well, not all of us are monsters," Zeena said sympathetically, relating to him.

"Thank you, stranger. For a moment, I thought I was going to die just because I stole bread to feed my kids," another prisoner spoke up.

"I second that. I can't believe I almost died because of the flawed justice system. I didn't even commit the crime; I was framed," said another prisoner.

"Me too. I don't deserve to die just because of some mistake I made years ago that I deeply regret," said a third prisoner.

Back at HQ, Alon made his way to the lab but on the way, he noticed the door to the cells open. He peeked inside to see Lyra's parents sitting alone in the cell with no trace of Lyra. Alon quickly moved past the room. He didn't want his emotions to get in the way of the mission, but his curiosity got the best of him. He kept thinking about what had happened to Lyra. Did her parents get to her, or did she escape? He shook his head, refocused on the urgent task, and increased his hover's speed to the lab. His mind was still in the cell, and he used every ounce of his will to stay focused.

At the inter-realm portal, everyone was gathering with as many survivors as they could find. There were some survivors who had been escorted there by the police. Amy made a record of all the people they had rescued while the rest waited for Alon.

"I can't believe Lyra did all this," Mathias thought, his mind stuck on Lyra, her innocent face. She had

been a pillar of support for him when he first learned about his parents' death from the Elder.

"Where is she anyway?" Zeena asked.

"She's gone. There were no traces of her left at HQ," Alon said as he handed over the missing part to Aria. Everyone looked at him, stunned by the information.

"Her parents?" Mathias asked.

"They are in bad condition, huddled in the corner of the cell," Alon replied. Then, looking at the many pairs of questioning eyes, he added, *"No, they haven't turned into skeletons or any other creature, if that's what you guys are wondering."*

They had lost more than they thought on this day, but dealing with their emotions would have to come later. Their people needed them, clear-headed and in their best form.

Aria finally fixed the portal and activated it. The portal was warming up, but the ignition caused a lot of noise, attracting hordes of skeletons towards the park where the portal was located. The portal finally opened, but the skeletons were getting closer. It they didn't evacuate faster, the skeletons would get to the remaining survivors. Alon flew back into the air on his hoverboard and blasted fire at the skeletons, creating a barrier between them while the survivors escaped. However, the skeletons were unfazed and casually walked right through the fire. The Champions got into position to fight the skeletons; however, the army of bones suddenly halted. The Champions took this as their chance to evacuate the last citizen and shut down the portal. This only left the Champions and the remaining skeletons. As the Champions turned back to rush toward the hordes for the final showdown, they noticed the skeletons had lost interest and were still frozen in place until they turned and started walking away, in one direction, in symphony—as if in a choreographed march. This left the team confused.

"Did we bore them?" asked Alon.

"Or they could be summoned," said Zeena. *"They are all heading in the same direction,"* she added.

"Well, then let's follow them," said Codeman.

*****

CHAPTER 6
KILL WHAT'S ALREADY DEAD

As the group followed the skeletons further, they saw more of them joining in, all marching in the same direction in complete synchronization. The scene was bizarre, piquing the Champions' interest. By the time they reached the outskirts of the town, the Champions had seen more skeletons than they could count.

Millions, if not billions, of skeletons had all gathered at Mount Gipfelus. The Champions decided to observe from a nearby outpost tower to see what would happen.

"Why did they stop?" Mathias thought out loud. All the skeletons had completely frozen in place, lacking any sort of movement.

"They might be waiting for something," said Aria.

"But what?" asked Alon.

The puzzle pieces fell into place when Satan arrived on the scene. As soon as Alon saw Satan, he was ready to jump him, but Zeena immediately used a spell that stopped all his momentum and froze him in midair.

"What's the rush?" asked Zeena.

"Whaddya mean, 'what's the rush'!" said Alon in frustration. *"And what was that for?"* he added. *"Shouldn't we wait around to see why he summoned the skeletons, to see his actual plan?"* said Zeena.

"I mean… I guess," stammered Alon, always ready to fight. Zeena released Alon from the spell, causing him to fall on his back. She caught him from falling and pulled him back up. Codeman and Aria looked at each other and rolled their eyes in unison before leaning on the railing as they waited for Satan to reveal his plans. Alon, embarrassed, looked around and saw Robbie, whose robotic eyes were rolling in circles with a goofy grin on his face. Alon glared at him and smacked him. Codeman snorted while Satan's voice boomed in the air, diverting everyone's attention toward him.

"Greetings, my subjects. I have summoned you all here to remind you of your duty. You are forever indebted to Him for giving you another chance at life, and so you shall serve Him to the end of time," Satan proclaimed.

"What the hell is he on about?" said Alon. *"And what's this 'IT'?"* he added.

"Perhaps he's referring to Zerlin's Stone," replied Codeman. *"Will you two shut it?"* said Neveah.

Satan continued his monologue with nothing much of significance until he mentioned the plan.

"Subjects, when He brought you back, He had a purpose for you. Underneath the EQ Haven, Champions' HQ, there is a library. This Library has years' worth of knowledge and precious artifacts," Satan said. *"And He wants us to retrieve this specific artifact,"* he added, pointing to a 3D projection of a weird artifact that looked as if it had been broken. It had cracks all around and basically no shape at all. *"Subjects, you have done me proud before, and I'm hoping you do it again,"* said Satan.

"You got all that on recording?" Codeman asked. Robbie, who was scanning and recording everything with his AI eye, replied with an affirmative nod.

"Snap a picture of the projection, time to get his ass," said Codeman as he jumped off the outpost and rushed to battle Satan alone.

"Why did he go off on his own?" asked Neveah, puzzled.

"We should worry more about helping him out," said Aria, and so the rest of the Champions followed.

Satan noticed Codeman flying towards him. Unamused and confident, Satan watched as Mathias, Codeman, came in, striking a punch that pushed Satan down. Satan rolled back to break his fall and got back on his feet. Now, Satan and Codeman were standing right opposite each other. Seeing the Champions approaching, Satan ordered the skeletons to take care of them while he dealt with Codeman alone.

Codeman walked closer. "*I killed you once, and I'll do it again,*" said Codeman.

"*I'd like to see you try, boy,*" Satan responded, spitting away a few loose teeth and blood that he lost from the punch.

The two ran towards each other at full speed. Codeman went in for another punch, but his arms went right through Satan. Satan healed his body and trapped Codeman's hand inside him. Now, without his hands, Mathias was unable to fight back. Satan picked him up and threw him at a boulder. Mathias' helmet shattered by the impact, knocking him unconscious.

Meanwhile, the team struggled to fight the city's worth of skeletons who were unbothered by their attacks. Anytime the Champions landed a hit, the skeletons would quickly regenerate. Completely outnumbered, they relied mostly on Aria's shadow clone army, but with the sun still up, the army was incredibly weak.

"*Neveah, can't you make them all go to sleep?*" yelled Amy as she reloaded her weapons.

Neveah tried to knock all the skeletons to sleep, but her spells failed as the skeletons' minds seemed empty with no conscience, preventing Neveah from taking control.

"I've tried, but it seems like they have no conscience," replied

Neveah.

"How does that make any sense? How are they able to move and make choices?" asked Alon.

"Beats me," Neveah replied.

The Champions, holding back and overwhelmed by the skeletons, were cornered and barely surviving.

When Codeman opened his eyes, he saw Satan in front of him. Satan picked up Codeman by the neck and slammed him back to the ground, further ripping away the armor. Codeman tried to sweep-kick Satan to make him lose balance and buy some time. But just like before, his legs went right through Satan's, and Satan instantly healed. Satan scoffed, seeing how pathetic Codeman was on the ground.

As Satan approached to deliver the final blow, he lifted Codeman's un-helmeted chin. *"Any last words?"* asked Satan.

Codeman then started laughing, his eyes beginning to glow purple. He rose as if all that beating had never happened. He clenched his hands, and a dark purple aura erupted from them. Satan recognized this look all too well—it was the same as when he had died. But with his newfound invincibility, he wasn't afraid.

Codeman punched Satan away, landing him on the ground, six feet away. The punch was so hard it shattered the rest of Codeman's armor, leaving him only in his hoodie and sweatpants. Slowly approaching Satan, Codeman put Satan in a choke with telekinesis, lifting him up without even touching him. While Satan was in the choke, Codeman's aura grew stronger, and his eyes went more purple to the point they were glowing, and all he had was a sinister grin on his face. Satan's body slowly started burning away, turning into ash.

From a distance, Zeena noticed what was happening with Satan and Mathias. She was all too familiar with that purple aura. She ran towards them, hoping it was not too late to recover Mathias. Zeena swung a rock at Mathias' head to knock him out, but the rock broke and turned into dust upon impact. She hesitated, unsure if she had the heart to do what was necessary. Then, with a bit of hesitation, Zeena held her hand up, her eyes glowing the same purple, and her aura started matching Mathias', but this time the target was not Satan. It was Mathias. Zeena began to seize Mathias's magical abilities, absorbing them, including the corruption. But Mathias, still corrupted and thirsty for power, attempted to use the same spell he used on Satan, on Zeena. But it was of no use. Zeena was more powerful and experienced in corrupted magic. Satan, seeing the situation before him, fled the scene before anything got worse.

Mathias fell to his knees as the last of his corrupted magic drained, and he returned to normal. His eyes returned to their original greyish blue, and the aura from his hand disappeared. Shaking his head, Mathias tried to grasp what was happening. He saw Zeena in front of him, floating with a purple aura all around her, her eyes now completely purple. Seeing Zeena, he was reminded of what had happened and realized how he might still be corrupted, and how Zeena had sacrificed her own rehabilitation by using the full power of her magic to save him before he went too far.

Moments later, the rest of the team caught up, the skeletons fleeing with Satan. The Champions returned to HQ. Although they didn't manage to capture Satan, they had valuable information about their next moves.

It was also fair to assume that these skeletons were Satan's army.

****

CHAPTER 7
THE HAUNTING PAST

The team gathered back at Headquarters to recollect themselves and to prepare for the oncoming heist that could happen. Zeena was still comatized after her sacrifice. Her eyes were glowing purple, and the veins on her hand were visible, glowing in the same purple. The team was worried about Zeena's state. She was not only knocked into a coma, but they feared whether she would be the same Zeena, when and if she woke up.

The team had brought Zeena to the Med-bay, where Amy laid her on a nursing bed. While everyone was gathered around her, Amy connected her with a device and turned on the medical scanner. This state-of-the-art medical scanner could not only execute multiple scans ranging from MRI to X-rays simultaneously, but the embedded AI technology could also assess all those scans and come up with a diagnosis, all in a fraction of a second. The scanner printed a report, and Amy finally declared that despite Zeena's body being in a coma, her brain still showed some activity. She further announced that her brain activity indicates that she is stuck in some recurring dream.

"If there's brain activity, can't you attempt to make contact?" asked Alon, scratching his head in confusion.

The whole team turned to Neveah with questioning eyes. Seeing everyone glaring at her, Neveah bashfully shrugged and said, *"I could always give it a try."* Despite her rivalry or hate towards her cousin, Neveah didn't want to see her die, especially because her experience was a huge asset to the team. And, after all, it was the heroic thing to do.

Neveah placed her hand on Zeena's forehead and attempted to read her mind. But Zeena's corruption would just block her out. Neveah tried again but the same would happen. Neveah then tried another spell, a projection spell that not only could let the spell caster read minds or see dreams but also become an active part of them while remaining conscious in the brain. Neveah put her hand on the forehead yet again. At first, she was blocked out again, but slowly, she started to get glimpses into Zeena's mind. However, as Neveah would see more into her mind, her eyes would glow a faint purple. Neveah snatched her hand back.

"It seems that when you start entering her brain, the corruption will start spreading onto you. And if my calculations are correct, judging from your attempt, you will only have an hour to save her," said Aria.

"Aria, you go research the library, and Alon, you guard the library and watch out for skeletons or Satan. Neveah, you get the job done," said the Codeman.

Everyone nodded and dispersed into whatever job they were given. *"Don't worry. We'll pull you away before it gets to the point where you are fully corrupted,"* said Mathias. Neveah nodded and slowly placed her palms on Zeena's forehead, and so the countdown had begun.

Neveah woke up in her old house, back where she lived with Zeena. Zeena had been orphaned at a young age as her parents were killed by Satan's disciples. Ever since then, Zeena has come to live with Neveah and her parents. Neveah saw a young version of herself running into the room, crying to her parents with a young Zeena following. It was obvious that none of them could notice Neveah's projection. Neveah was all but a spectator of Zeena's recurring nightmare.

She heard young Neveah complaining about how Zeena had been drawing on her journal. Zeena had always been outcasted by her cousin. Whenever she would try to hangout, Neveah would shut her off. The loneliness and isolation had followed her into her middle school days. Neveah saw as the two young versions of her and her cousin ran off.

The scene around her changed into her middle school, and she heard yelling from behind. It was yet again Zeena arguing with Neveah and her friends. And this trend would continue with every new scene. Neveah could see how Zeena was affected by these arguments. The whole school thought of Zeena as some psycho due to her dark nature and constant arguments with everyone. Zeena would become more isolated and colder towards others. Neveah realized that the entire time, she assumed Zeena was some bully who hated happiness; in reality, she herself was the bully. She was the one who made Zeena feel outcasted both in and out of her household. Now, in each memory, Zeena would become colder and more sarcastic; she would avoid talking to anyone, and anytime someone would try to talk to her, she would be rude and cold and turn them away.

The scene then again faded into Zeena and Neveah's high school days. Neveah looked around the place and immediately recognized it. The day seemed normal, with Neveah and Zeena walking to school. The two would go walk in complete silence not bothering to commute because they both knew it would end in an argument. Neveah walked with the two high schoolers, and she kept looking past her shoulder as if she was checking for something. And then she saw him, it was a guy in a full black hoodie and black cargo pants peeking from an alleyway. The guy had been following the two girls for a while now, stalking them at every turn. It was awfully eerie seeing him trail behind, that was until he charged at the two. The guy grabbed Neveah, knocked her unconscious, and ran away with her. Zeena was slow to react as she couldn't hear anything from the music she was playing. However, she knew exactly who had taken her cousin away. Neveah had recently turned down a senior who had confessed his feelings to her. And this boy could not deal with rejection. Zeena would often see him staring at Neveah across the lunch table while he would cut his food. And what further confirmed Zeena's hunch was that the guy was about the same build. But despite knowing all this, Zeena had kept quiet. After all, why would she want her bratty cousin back, all she would do is make her life worse. For all Zeena knew, the absence of Neveah would have been better for her. But she was swiftly proven wrong.

The evening she got back home, Neveah's parents had already found out about the abduction; they blamed it all on Zeena for not doing anything while her cousin was taken away. Zeena was grounded. And so, in hopes of improving her situation, Zeena sneaked out to try to find Neveah. Zeena managed to find the senior's address in the school records, and as she left school, she headed straight to his house. Zeena approached the front porch and pressed her ears against the front door. After she confirmed she was clear, she pulled out her hairclip and started picking the lock. The door swung open, and Zeena tip-toed her way into the kitchen; there, she grabbed a knife for protection, just like she had seen in old movies, and attempted to look for Neveah. As she entered the basement, she peeked below the staircase and saw Neveah tied up to a chair with the stalker being on a phone call. Zeena took this as a chance and ran charging towards him, tackling him to the ground and repeatedly stabbing him. With each stab, she let out more of her past traumas. All until the stalker was deadly injured. Zeena then walked to Neveah and untied her, and the two walked back to their house. Soon enough, the consequences of Zeena's actions followed. There had been investigations into the assault. According to the authorities, Zeena should have contacted the police with the information. The act of such violence was completely unnecessary and required reprimanding. The police showed up at their doorstep the following day with a warrant to arrest Zeena. Soon enough a trial was held. And with all the evidence gathered, Zeena was most likely going to juvie with a court trial pending till she was a legal adult. That was all until a new form of evidence had been found, a corrupted spell book found under the floorboards of her room. Zeena was granted 6 years of rehabilitation, and if no progress was found, another trial would be held to decide her future.

Now the scene changed to Zeena's rehab. She was chained up and there were shackled around her ankles. Thousands of guards at post, all treating Zeena like a monster. Every week on a Thursday, a therapist would visit her, and would record her progress and that was all she would get. The holding facility was a depressing sight and Neveah could not imagine staying there for 6 years. She didn't know if she could last even a day. Neveah could not believe how she had treated Zeena this whole time. Zeena was always trying to fit in and all that Neveah did was shut her out. If anything, Neveah was and would record her progress of spending six years in hell. And even then Zeena believed she was in the wrong and apologized to Neveah. If it weren't for Zeena, who knows what would have happened to Neveah.

Now the scene cut into an empty void. Zeena was in those same chains and shackles crying as she had no hope for escape. Neveah walked up to her. Zeena looked up at Neveah and burst into even more tears, apologizing for all her actions.

"What are you apologizing for, if anything I should be the one apologizing. I treated you like absolute shit, always outcasted you. I'm the reason you were hated in school, I'm the reason you turned to corrupted magic. I was supposed to be the sister you never had, but instead I was a total jerk. And even after you came back, I couldn't let go of the past, even if you had saved my life that day.". The two then hugged, finally letting go of their grudges, and with that the shackles around Zeena had broken, symbolizing her release from corrupted magic.

In the real world, Amy noticed Zeena's eyes turning white and the glow in her veins going away. Her body was starting to recover as well and it seemed that she was out of her coma.

"*She did it, Neveah did it,*" said Amy. The two then woke back up, Zeena's eyes glowed white and the returned back to normal. "*Thanks, Nev*" said Zeena.

"*We have a lot to talk Zee,*" Neveah said with tears in her eyes.

"*Making up for the lost time?*" Zeena replied in a weak but happy voice.

Mathias didn't know the exact situation but he somehow had the feeling that the two sisters need time with each other. He said clapping his hands, "*All is well that ends well. Zeena you rest, and Neveah you look after her, Amy you follow me to the Library.*" He winked at the girls and exited the room.

****

CHAPTER 8
MORE ON NECROMANCY

Codeman entered the ancient library with Amy following behind. Aria was scanning books with a Book Synopsis Extractor. This was an AI-integrated device that would scan a seal on the cover of ancient books. The seal would contain codes that the device could decode and, through that, translate and summarize the main content of the book. There were thousands and thousands of piles of already scanned books, and the two had yet to find anything regarding their situation or anything on the artifact. Codeman joined the two on their search and after emptying half the shelves of the library, they finally found a book that could possibly solve their problems.

"*An Account of the Great Apocalypse, by Zerlin himself*," Aria read from the Extractor as she raised up the book, showing it to the rest.

Mathias, the Codeman, could recall when Robbie had narrated the story of Zerlin's stone. He mentioned Zerlin being frustrated by the murder of his family; hence, he released an army of the undead on the town. This could help them find out how to defeat Satan in his new dead-yet-alive state. Aria placed the book on a table and the three gathered around it. Aria opened the first page which left the three completely disappointed. The writing was all smudged, and the pages were flaking to the point where they could crumble into dust with only the flick of one page.

Are you kidding me? What are we to do now?" Said Codeman in frustration. *"It tells how much more we need to and can invent with the power of AI. Right now, we don't have anything to restore this book."* he mused.

Aria put in her own thought, *"You are right about the fact that we have not yet leveraged the full potential of AI advancement. A lot can be done, and a lot is in my mind, too. Only if we soon get done with solving these Satanic evildoings."*

"Anyways, there is one more way to find out what happened in Zerlin's legend," said Aria. *"The BS Extractor does state that Zerlin belonged to a village called Zalzar in Netherheim,"* she added.

Mathias paused for a second before cooking up a plan. *"You, Amy and Alon stay here and guard the library, try to find more information and take care of Zeena. I, Robbie, Trophy, and Neveah will visit Zalzar to find out about this Apocalypse,"* Codeman took the lead and conveyed his plan to Aria.

"OK, let me know what armor you require for this trip. I will get them arranged as soon as possible?" Aria asked.

"Nah, there's something I have been working on," the Codeman replied. He asked Robbie and Neveah to meet him at his lab and to bring Trophy. The four gathered at the door of the lab. Mathias entered his fingerprints on the door, and the door opened to reveal a team of AI robots working on a new armor and a clone of Codeman directing the operation.

"You were a cloned all along?" asked Neveah.

Neat, isn't it? But only where my clone presence can do the task. I trained in clone magic and improved clone strength, duration, and proximity. Now, I can have a clone on the other side of the realm with no issue and retain my original strength, and the best part is, the clone can live on forever," Codeman explained.

Codeman also explained how this new armor was not only able to achieve everything that his previous armor could do but it was also made with a PlathmatiteTitanium alloy, making it nearly indestructible. The armor also had a new integrated AI called MAHIA (Miniature AI Human Intelligent Assistant). Apart from helping the Codeman assess situations, this AI assistant could pilot the armor when Mathias was not wearing it. It could use spells without Mathias having to conjure them or him being present and MAHIA could detach from the suit and fly around acting as a humanoid spy drone.

"Isn't this what you were telling Aunt Billie about, the invention we were about to exhibit in Techcon the day you were abducted?" Robbie Interrupted.

"Precisely, MAHIA has been a passion project of mine for as long as I can remember," said Mathias with his voice filled with excitement. But then he saddened as he added, *"When Aunt Billie used to tell me stories of Peter Pan and Tinkerbell, the engineer who met a fairy, and the two would work together to combine their abilities to fight pirates. I felt inspired to do something similar myself. So, I created*
MAHIA."

"She would have been proud of you," said Robbie, comforting Mathias.

I can only hope so," said Mathias. *"Thanks, Robbie,"* he added as he moved forward to continue showing his inventions.

"And this is another invention that I have wanted to make for a while, but the technology was difficult to achieve without the aid of powerful magic," said Mathias, pointing towards a small armored dog harness. The harness was able to safeguard any animal wearing it, with functions such as fireproof materials, being able to float on water, accelerating running and swimming speed, and even enhancing the senses of the animal. But perhaps the neatest thing about this new piece of technology was the built-in AATD, the Animal Audio Translation Device. This device could pick

up an animal's voice, and by checking the tone of the voice, assessing the environment, and reading brain activity with the help of Artificial Intelligence, it could provide a rough translation of the Animal's voice. It would seem that the animal or pet is talking to you in the human voice. Mathias went on a monologue, reminiscing the times he had with Trophy, "*I still remember the day I rescued Trophy; He was being attacked by some big dogs and had lost half his limbs and even parts of his head. It was a miracle that he could still survive. That day me and Aunt Billie worked tirelessly so we could make Trophy cyborg parts to give him a chance of survival. That day, I managed to save him, and now he saves me; whether it's Satan's dungeon or that room in the Ancient Library, he always jumps into danger to protect me, so why not allow him to participate in missions with us rather than having him stay in the background.*"

Mathias did not look like the Codeman at this moment but a vulnerable young man. He wiped a tear off his cheek and wrapped the harness around Trophy. Everyone in the room looked towards Trophy, waiting to see if the AATD had worked, and then Trophy spoke, "*Thank you!*" And that's all he said. Tears ran down Mathias' cheeks as he hugged his companion after all they had been through.

Mathias recollected himself. "Enough chitter chatter. We have a mission on our hands," said the Codeman in a bold tone. Mathias equipped the armor and entered the coordinates for Netherheim on his remote inter-realm teleporter. Instantly, the four vanished into thin air and reappeared in Netherheim.

"I don't think I can ever get used to this place," said Robbie, looking around the shady alleyways and the dark nature of the Netherheim streets.

"You and me both, buddy," said Mathias, *"But we are here on a mission, and we need to find Zalzar."*

Just then, a strange old lady noticed the group mention Zalzar and with her interest piqued, she approached the Champions. *"Are you young'uns lookin' for Zalzar,"* she said in her scuffed voice.

"Yes, ma'am, could you perhaps guide us to the village? We are the Champions of the Realms. You might have heard of me, I'm Codeman. I saved the realms from Satan," said Mathias.

The woman gave Mathias a strange and confused look before telling them to follow her.

"I don't think she knows you," whispered Neveah to Mathias, poking fun at him.

"Yeah, I figured," said Mathias, rolling his eyes at Neveah.

"That was pretty embarrassing if I do say so myself. I don't know how I'd recover," cackled Robbie having the time of his life teasing Mathias.

"I thought you were on my side," said Mathias, and the group moved forward with chit-chat and banters, pulling each other's legs. After all, despite the heavy responsibility on their shoulders, they were still young.

****

Back at the Ancient Library, Aria and Amy were still searching for information regarding the apocalypse, hoping to find something new. They also hoped to find the artifact that Satan had shown in the projection, as he claimed it was somewhere right in these walls. While scouring bookshelves, Amy found a spell book regarding the magic of healing or regeneration.

"How did Mathias manage to perform magic, weren't only the people of Netherheim and EQ Haven able to perform it?" Amy asked Aria.

Aria looked up from the book she was scanning through, *"I don't know what misconception there is among the people of Techtopia regarding magic because I can swear Mathias said the exact same thing,"* Aria responded.

Amy's eyes widened, and with excitement in her voice, she exclaimed, *"Is that a yes? Can I perform magic as well?"*

Aria gave her an affirmative nod in her usual 'talk less, do more' way before going back to reading. Ever since Amy had joined the Champions, she was mostly an onsite doctor for any injured and was barely able to help in battle without any weapons. Even when caring for the injured, she would barely have any resources to work with in chaotic situations and would usually have to work with scraps. Being able to use magic could help her be more helpful. Aria noticed Amy was overjoyed by the concept of her performing magic, so she offered to investigate the rest of the library herself while Amy could try out a few spells. Amy agreed and ran out of the room with some books under her arms to practice some spells.

Meanwhile, at Netherheim, the group had been walking for hours. "*Are we there yet?*" asked Mathias, but the lady didn't respond. The group became exhausted from all the walking, but the lady didn't break a sweat and kept on moving.

Why are we not using our hover and jet shoes? Whispered Robbie, noticing the exhaustion on Neveah's face.

"*Because we don't want to overwhelm the lady. She looks old school,*" the Codeman replied.

"*Yes, very very old school. Thousands of years old,*" Robbie gave a wide, toothless grin, and Mathias ground his teeth at him. Robbie was an AI Robot. No wonder the matter was funny to him. Exhaustion was unknown to his system, unless some spare part malfunctions or his battery drained out, which Mathias so wished to happen at this time. He thought of taking a taser to his system so he could feel how they were feeling. The thought brought an evil goofy smile on his face.

Robbie raised his robotic hands and popped a finger crackling with electric current, "*Don't you dare go there. I can see the evil in your eyes. And I am prepared.*" He raised the second hand, popping another finger.

Mathias sobered up seeing the current flaring out of the steel fingers, didn't want to risk his well-being at this crucial time after all.

"*I don't trust her; I'm getting weird signals from her brain,*" said Neveah.

"*Not like we have a choice. Zalzar is not on any update of the Netherheim map that I can find across my database. The village must be ancient,*" said Robbie, putting down his hands and glaring at Mathias who was still giving him side glances.

"*Wasn't the apocalypse around the second Medieval era?*" asked Neveah.

"*Yeah, but the second Medieval era or the Modern Medieval era was known for Data theft, spyware, and deleting data,*" the Codeman explained. "*I bet MAHIA could recover some information, but she's still collecting data from the beginning of the universe up until now,*" he added.

"*Can we stop saying Data now, I just feel like this whole thing is a trap,*" said Neveah. Trophy gave an affirmative bark with the AATD following up by saying *"Agreed"* in its usual robotic voice.

"You should sell that stuff. I bet we could get a lot of funds for our projects," said Neveah.

"We're here," said the lady, interrupting the group's conversation.

"Finally," groaned Neveah.

The group saw an old, torn-down village right before them. The village was wreaked with the smell of corpses. The graveyard was probably bigger than the rest of the village itself, and according to the lady, there were more corpses buried under rubble, roads, buildings, farms, and even parks.

"Come on, I'll show ya young'uns around," said the lady.

"Thank you so much, but we are only here to investigate someone who goes by the name of Zerlin," Neveah answered skeptically.

"Well, then let's start there," said the lady. The lady guided the group to one of the bigger houses in the village. It seemed like Zerlin had earned quite a few with his scams. The lady unlocked the door to the house. The team walked inside the gloomy rooms of the property.

"Where exactly could Zerlin have hidden his research?" the Codeman asked the lady, but when he looked back, there was no one there, and then he

heard the sound of the ancient door locking behind them.

"*Why would she-*" said Robbie before he was interrupted by Neveah shrieking, "*Skeletons.*"

The group backed up into a corner with a family of weirdly shaped skeletons surrounding them. The Codeman aimed at them and shot with everything he had but nothing worked. Robbie departed from the group, leaving them alone with the skeletal family, and came back with a large blanket and his disappearing mode switched on. The bony figures seemed unbothered by Robbie as he was not a living being with flesh and blood but rather a metal robot invisible to them. This allowed Robbie to wrap the blanket around them, tying them up. Neveah, Codeman, Robbie, and Trophy busted the door down and ran out of the house, chasing after the lady. Neveah used her powers to put the lady, who was just a few feet running away from them, to sleep. And when she woke back up, she was at Zerlin's, tied up to a chair.

"*Lady, what the hell. What did you do that for,*" asked Robbie.

"*I'm sorry, young'uns, but if I were given a chance to go back, I'd do it again. I'd do anything for my family,*" the lady said with determination.

Robbie did a DNA scan on the lady and concluded that she was Zerlin's daughter, Zoey. "*Wait, so you are Zerlin's daughter?*" asked Codeman.

The lady nodded. *"How are you still alive? Didn't the mob get to you? If not that, then surely by old age?"* asked Neveah.

"First off, very rude. Us wizards and witches live centuries-long; some would say I'm in my golden age. About the mob, when the people of the village were murderin my family, I was out to study. When I came back, this place was flooded with what you call dem undead." Zoey replied.

"Well, why betray us?" the Codeman followed up.

"I can't let my family starve now, can I?" Zoey replied. *"Now can ya young'uns let me go already?"* She added as she grew frustrated.

With all the confusion that had gone on, the group had finally remembered what they had come here for and asked.

"One last question: how? How did the apocalypse end?" the Codeman asked.

"Now, why do I tell ya young'uns that? What do I get for it? All ya young'uns did for me is waste my family's meal," said Zoey.

"Please, EQ Haven is going through the same curse, Satan. This malevolent villain is part now in a zombified state and is terrorizing our realm. we need to stop it. People are dying and are losing their homes," pleaded Neveah.

"*Not ma realm, not ma problem,*" Zoey replied, shaking her head in denial.

"*What if we bring them back? What If we bring back your family?*" said Codeman as a last resort.

"*Mathias, we don't even know if that's possible. Don't you think it's a bit far to promise th-*" Neveah said before she was interrupted by Zoey. "*Deal!*" said Zoey. "*You wanna know how the apocalypse was solved? Well, they obviously burnt em. Burnt it all down to the ground. If there's one thing these bones hate, it's fire. There's also dem corruption curses and whatnot, but that could kill anyone if you're being realistic.*" Zoey replied.

"*Thanks, ma'am. You don't know how many lives you have saved,*" said the Codeman as he untied her. The Codman pressed his earpiece on his helmet, "*Aria, are you there?*" But they had no reply, and all that could hear back was the sound of the intruder alarm at the library. They looked at each other with puzzlement and fear in their eyes.

****

CHAPTER 9
FAMILIAR OF FRIGHT

Just a few moments earlier, Aria had finally found a book regarding the fragment Satan was talking about. As soon as she sat down to read it, she heard loud thuds and then Alon's grunts and shouts, as if he was in trouble. Aria quickly used her magic to hide the book in darkness and rushed to aid Alon. As soon as she got up and ran out the door, something threw Alon back at her, making them both fall. Then, Satan walked in with his skeletal gladiators. The sound of the alarm echoed through the hallways of the Champion's Headquarters and was heard by Amy and Zeena. The two rushed towards the library to see what was going on. When Aria and Alon crashed and fell, Aria's earpiece had bounced out. Through the earpiece, she could hear Mathias shouting out for anyone who was there.

Before anyone could utter a word, Satan stomped on the earpiece and calmly asked, *"Where is it?"*

The group knew exactly what Satan was talking about. He was here for the artifact.

"We don't know where it is, and even if we did, we'd never tell you," said Alon with a snort.

"*Then die you shall,*" Satan responded, and with that, hell broke loose, and a battle began. Alon, Amy, Aria, and Zeena assembled and stood head-up in front of Satan and his army, without a clue but determined to fight till their last breath.

While the library was under attack, the group back at Netherheim was desperately trying to contact the team. The Codeman knew something was up; the alarm and the lack of response must have meant Satan was there. Mathias entered the location for the teleporter as fast as he could so they could help the rest of their team. But something seemed to be interfering with their signals, and they couldn't get back home.

"*Damnit, this thing is busted,*" said Codeman, as he repeatedly hit the teleporter on his gauntlet.

"*What now? Do we just leave them to fight this one on their own?*" asked Neveah.

"*No, I think I have a backup plan,*" said Mathias.

"*Remember the clone I had in the lab? If I can manage to create a psychic connection with him, we might be able to warn the others of how to deal with this dead or undead army,*" Codeman added.

"*How do you plan on making a connection from another realm? Sure, you might be able to sustain the clone, but having a psychic connection to let him think what you are thinking and improvise is a whole different level,*" said Neveah.

"I can try," said Neveah nervously, doubting her abilities but pleased by Mathias' trust in her.

"You have to. It's perhaps our only hope," said Codeman.

The two held each other's hands, attempting to make a connection with the clone. For a moment, Mathias could feel like his consciousness was able to make contact, but the connection wasn't strong enough. The connection seemed to be weak because Neveah's head was split, and she was distracted.

"Neveah, you need to focus on establishing the connection. Get your head in the game," Codeman firmly tried to get Neveah's attention, not knowing that every fiber of her being was focused on Mathias and their joined hands instead of making the connection with the clone.

Neveah gave an unseen slap to her head and refocused on the important task. And so, the two reattempted to make contact, and with Neveah now concentrating her energy, the connection was made. Codeman's clone started getting a huge headache. He fell to the ground as the pain in his head grew more and more, and he started getting flashbacks. These flashbacks showed the clone everything that had happened in Netherheim, alerting him of the weakness of the skeletal army and the possible intrusion in the library. With the clone being all caught up, he equipped an older armor from the lab's armory and ran towards the library, which was now a battlefield. As he exited the soundproof walls of the lab, he could finally hear the alarm bursting through the hallways of HQ. Mathias' clone made his way down to the library to see the Champions backed into a corner, fighting the skeletons while Satan trashed the place, looking for the artifact.

"*Guys, the weakness, it's fire,*" shouted the clone as he entered the scene. Satan turned around to see the Codeman facing him.

"*And here I thought you cowered and ran,*" said Satan.

"*Wishful thinking for you, isn't it?*" replied Codeman. Codeman activated the flamethrower mode in his suit and aimed toward Satan. As the clone blasted the flames towards him, Satan jumped out of the way and used some skeletons as a meat shield—ironic. Meanwhile, the Champions who were backed into the corner started fighting back. Alon decided to fight the skeletons that would come near them and slow them down while the rest tried to look for the artifact before it got into the wrong hands.

As Codeman burned away more of Satan's cover, his fuel for the flamethrower began to run out. In mere seconds, he would again be powerless against Satan. With the Champions slowly losing their upper hand, it wouldn't take long before they were overwhelmed. The team was running out of options, so as an act of desperation, Aria yelled at Alon to use the same move that he had made against Glitch at the park.

"But that would burn everything in the library, all that information, history, and knowledge that has been preserved for centuries, all wasted," yelled Alon.

"That includes Satan. We don't have any other options. It's the only way," Aria yelled back as she fought off more hordes of skeletons. Alon knew what he had to do, so he charged up his ability while the Champions took cover, and then he released fire in all four directions. The flame filled up the room, erupting in every direction, burning every shelf, every door, every wall, and every book. With the whole library burning down and all the metals melting with the extreme heat, the library was about to collapse. However, before the pillars of the once ancient and preserved place fell, the artifact was exposed behind some burning rubble of a fake wall. Satan came out of hiding and ran towards the artifact. The Champions stood in his way, but with no flamethrowers and Alon's energy being drained, they were left with limited options. Zeena stepped forward. Although she had barely made it out of the corruption the last time, she had to attempt this. The situation was too urgent. Zeena's feet lifted up as she began to float. She was using corrupted magic again, but instead of her eyes turning purple and a purple aura erupting from her arms, it was pure white instead. Zeena was ready to succumb to the darkness that would come with using corrupted magic, but this time,

it was different. There were no negative emotions that overwhelmed her. She felt like she was in control and she felt like she was free. Effortlessly, she lifted Satan into the air and slowly encased him in stone, a stone he wouldn't be able to escape. Everyone watched in fascination and awe as Zeena's feet touched back to the ground. She was still surprised at what she had achieved. Instead of the corrupt magic taking hold of her mind, it was she who was controlling the game. The team broke out from the trance and began to celebrate what they thought was a clear victory.

Just when Alon was trying to hook Aria into giving him a high five, a dark mist started to approach and covered the ground. Dark clouds roofed the ceiling despite the indoor environment, and a dark aura filled what was left of the library. Strong winds blew inside, extinguishing the flames and turning the library into a cold dark hellscape. Glowing purple eyes approached the group. As the figure got closer, Codeman, who was connected mentally to his clone, could recognize the entity—The hooded figure with a goat's skull for a head with large horns and fangs. Its eyes were a void with only a faint purple light shining out of them. It hovered over the mist as it got closer. It held a staff with a dark purple cloudy crystal floating atop. Then it finally clicked to Codeman. This was Satan's right-hand man. This is the one who had killed all the Crusaders in battle. This creature was the one who had defeated Codeman, and yet no one had believed in his existence. Mathias had thought he had gone crazy; he was treated as if he was crazy, but now everyone saw him. As Grim got close, the Champions started getting dizzy and fell unconscious, and when they woke back up, the artifact was gone.

Back at Netherheim, Mathias could feel the fatigue, the haziness, and the fatigue that his clone had felt. Robbie noticed Mathias going quiet.

"You okay, buddy?" asked Robbie.

"It's real, and it's here," mumbled Mathias before falling unconscious. As Mathias collapsed to the ground, his Gauntlet finally reached signals and was ready to teleport them back. Trophy rushed to the unconscious Mathias and started licking him, trying to make sure he was alright.

"What the hell is he on about?" said Robbie.

Trophy barked, and his translator followed with saying, *"Beats me."*

Robbie glanced at Trophy, sporting an oversized helmet, *"You sure that thing's gonna fit through the teleporter with that ridiculous helmet?"*

Trophy barked back, *"Better safe than sorry. What if there's a space vacuum in the teleporter?"*

Chuckling, Robbie shook his head. *"Right, because the first thing I worry about with interdimensional travel is running into a vacuum cleaner."*

With a playful growl and a wag of his tail, Trophy seemed to laugh in his own way. *"Alright, space dog, let's get back to base. Helmets included,"* Robbie said, giving Trophy a pat on the head. The brief moment of levity was a welcome distraction from the tension they were all feeling.

"*Give me a hand, let's get back to base,*" said Robbie, eying the Gauntlet, blinking a green light, ready for teleportation. Trophy gave an affirmative bark and started expanding his bionic limbs, holding Mathias up. Robbie pressed the button on the Gauntlet on Mathias's arm and used the teleporter to return to the library. When they got back, the whole team was on the ground. Their eyes were empty voids, and they were trapped in their own minds, just like how they had found Mathias when he had first encountered The Grim.

Neveah went ahead and woke everyone up one by one. "*Rise and shine, sleeping beauties... Except Alon,*" said Robbie, clapping his metal arms around the room as Trophy followed behind, barking to wake everyone up.

They seemed traumatized. But no one seemed more afraid than Mathias, as he had witnessed Grim's true capabilities. The Champions woke up and gradually got back to their senses with the help of Nevaeh's capabilities.

Once back in the world, Alon shrieked, "*The artifact is gone.*"

Aria groaned in a weak voice, "*We can see that, Alon.*"

Everyone noticed that the artifact and Satan encased in the stone were both missing. And with Mathias being in no state to fight, the team had to retreat and regather. But Aria had one thing in her hold that could show them the way forward: the book she had found and hidden in the library. It was time to retrieve it and research what they had encountered and what they were after.

Nobody knew that soon the present would collide with the past. They will return to the place, the origination of humanity, the beginning and the end too, where it all started—the Earth.

****

CHAPTER 10
MOTIVES

Aria placed the book on the table, and the Champions gathered around her in hopes that the book would hold the answers to their questions. They were all curious about what was going on, but Mathias was still scared. He had seen Grim effortlessly take down an army of Crusaders and the Deathbringer in a matter of seconds. In his second encounter, Grim had defeated a whole team of superheroes. Mathias doubted their chances of defeating Grim; he didn't even know where to start.

"You guys don't understand. We can't defeat it," said the Codeman.

"There's no such thing as never," said Alon, trying to reassure Mathias.

At this moment, Mathias wasn't the Codeman, a superhero. Instead, he was a scared lad who feared for his friends' well-being. *"Did you see the same thing that I did?"* exclaimed Mathias.

"Bro, after the countless times we had our butts handed to us, this one is the most humiliating. Remember when that one clown with a machine gun

soloed us in front of the whole town? I didn't want to be seen with you guys anymore. But the point is, we got him back, we always do," said Robbie. Trophy giggled, and Mathias grunted and glared at them, trying to eat them up with his eyes, steel and all, for bringing up their stupid adventures and tainting his Codeman persona. *"Really, what happened? I want details,"* Zeena asked, getting interested in the comic scene Robbie was painting, but Aria intervened.

"Mathias, what has gotten into you? You are a leader who fears nothing, and now you're going to cower without even trying to find out what's happening," said Aria.

Mathias was still afraid, but deep down inside, he knew Aria was right. He nodded, and Aria blew the dust from the cover, revealing the title, 'Realm of Gods.' Aria opened the first page and began reading.

Long before the Big Bang, there was a Supreme God, the all-knowing and all-powerful being. This God created a world for him to inhabit, crafting everything with a purpose. Soon enough, the Supreme God birthed more gods to look after and manage his creations. There was a God for each and every element, whether it be Fire, Wind, Water, Earth, Cosmos, Time, Life, or whatever else existed. As time passed, these gods created slaves, half biological and half mechanical beings, called angels. Angels were built to worship and serve these gods without free will, or so it was thought. The gods had made a vast and enormous Universe with very little to inhabit it, so they decided to add more life in the form of Fauna. They first tested the waters with the creation of dinosaurs and Jurassic animals, but after that experiment failed, the gods decided to spawn a new creature, a new species, Homo sapiens - beings with full free will, primates with the ability to adapt and evolve. The Homo Sapiens used to worship the gods and turn to them for any of their needs. But as they evolved, they became more creative, more independent, and less reliant on the gods, moving on and away from them. Most of the gods went into hiding as they no longer needed to overlook the Universe. They were no longer needed. This is when Grim, a powerful angel, realized that gods and angels were no longer in power. He noticed that humans, now called humans, were turning away from god.

Disgusted by life, Grim sought to destroy it. And thus, with this realization, he soon unlocked free will, defying the rules set by God. Grim started his mission by turning humans against each other. He initiated Corruption spells, making humans become negative. It didn't take long for the gods to catch on and banish him from the Realm of gods. Soon, Satan followed in Grim's footsteps, and he too was banished into the digital realm for trying to become an equal to the God.

"So how does that artifact tie into this story?" asked Amy. *"Maybe it's some glorified backscratcher that the*

gods had. You never know. Maybe Satan just has a spot he needs to itch," said Robbie. Robbie and Alon rolled over, laughing at the joke, and raised their hands for a high five but stopped their hands and laughter midway, seeing Codeman and Aria glaring daggers at them.

"*Does the book say anything else?*" asked Mathias. Aria skimmed the pages until she landed on a chapter titled 'The Portal,' with an image of the artifact scribbled along the side. Aria began reading.

When the gods felt that humanity no longer needed them, they fled to the Realm of Gods. Most of them wished to be forgotten, while some still left remnants of their existence. The Gods constructed a Portal hidden away in an unexplored area of a planet where they first sent all humans, The Earth. A planet covered in 70% ocean, once dominated by humans, is now left mainly abandoned as their descendants left in search of a greater, better home. However, there are rumors that the gods had left fragments of a key - A key that could activate the portal to the realm of gods. A key for the humans to find when they once again needed the aid and help of the Gods.

As Aria put down the book, a page fell out. It was a map written in an unfamiliar language with coordinates in an unfamiliar format. However, one thing that the team did recognize was the planetary coordinates directing to Earth.

"We are heading to Earth," the Codeman said boldly as he tried to suppress his fear and reclaim his leadership role.

The team headed to Earth, hoping to find answers before it was too late. Without a teleportation chamber or a portal to connect to, the Champions had to embark on an old-fashioned road trip. The team headed to the garage and saw the dusty old ship sitting at the dock.

"Oh, how I missed this beauty," said the Codeman as he approached his old ship. Alon fueled the ship up, and the team got ready to leave. The traditional method was extremely slow, giving the Champions time to cook up a plan and train for the bigger fight. Zeena tried adapting to her new abilities, as did Amy with her healing magic.

Days had passed, and the team slowly started to realize that all this was familiar.

"Is it déjà vu, or have we been to this part of the universe before," said Robbie.

"I can bet this is the place we came to when we got stranded that one time," said Mathias, recalling the time they were shot down by Satan and lost their fuel. The ship landed in the same jungle where the Champions had found the temple and the fuel.

"Time to finally test MAHIA," said the Codeman. *"MAHIA...activate,"* he added.

A voice came out of the Codeman's suit, *"Initializing MAHIA...Welcome, Codeman... I am MAHIA, your new personal AI Assistant."*

"MAHIA, run a scan of the planet and locate any form of advanced civilization," said Codeman. *"Commencing scan,"* the AI assistant responded.

Codeman loaded up his AR holographic screen with a full world map, and the pinpoints of signs of life were revealed in a matter of seconds. Most of Earth seemed fairly unpopulated and covered with overgrown forests and mountains, with some abandoned cities. However, there were three main locations spread out across different enormous islands. The Champions took out their hoverbikes from the ship's garage and made their way to the closest location.

Upon arrival, the Champions saw huge buildings and skyscrapers covering the ground. Each building was covered in neon lighting and text much familiar to the one on the map. The Champions entered the town and noticed the crowd staring at them and giving them estranged looks. The Codeman knew there was no point in hiding, so he bumped up the volume of his voice through the microphone on his helmet and shouted, "Inhabitants of Earth... We are the Champions of the realms. A huge danger approaches us, one that seeks to destroy all life. We need your help," He took out the map from his pocket and showed it to the citizens. "This… This is a map written in your language, and we need your help to decode it. May you please take us to your leader." He finished his speech, and as he looked back at the crowd, his optimism turned into pure disappointment as he saw the crowd looking confused as hell.

"Buddy, I don't think they understand you," said Alon.

"Yeah, so I assumed," Coleman said in a disappointed tone.

"You looked properly stupid doing that as well," Robbie added as if he was revealing a great wisdom. Mathias glared at him- no use. Robbie was least bothered.

Soon, some men came by and cuffed the Champions. The team tried resisting, but Codeman told them not to. *"They'll probably take us to someone we can talk to."*

The law enforcement escorted the Champions to a huge building in the middle of the city. There, they met a strange woman wearing clothes they had never seen.

"Kon'nichiwa, watashinonamaeha Naomi," said the woman. The Champions gave her looks of confusion.

Robbie leaned towards Trophy, *"Uh, Trophy, can you loan her your AATD (Animal Audio Translation Device)?"* Mathias kicked Robbie from behind. What if she understood they were thinking of using an animal's device on her? She could've taken it as an insult.

The woman rolled her eyes and grabbed a necklace from her pocket. She cleared her throat and said, *"Hello, my name is Naomi. What brings you to Asia?"*

"How did you… even we have just begun exploring this technology," said Alon in amusement.

"Well, over at this side of the planet, we are known for having technology for every small task," the woman replied.

"We appreciate and honor your advancement. But right now, we are in a very critical situation…" Codeman went on to explain their situation.

Naomi, though skeptical at first, chose to believe Mathias. *"Centuries ago, when humankind abandoned earth, most of us abandoned god. They became mere myths. Believing you is a huge gamble on our time and resources. But you must have come this far for a reason, and if what you say is correct, then life everywhere could be in extreme danger,"* said Naomi. *"I will assign you one of our people, Adamina. She is the smartest kid in town. I believe she can guide you."* Naomi summoned Adamina.

Moments later, a girl walked in holding a tablet. *"You summoned me, President Naomi?"* said Adamina. She looked up from her tablet and locked eyes with

Mathias. Mathias felt like pins were stabbing him from all around. He suddenly started sweating and looked all nervous.

"*Oh, brother, not again,*" said Robbie, rolling his eyes.

"*Simp,*" came a laughing voice from Trophy. But Mathias didn't respond. He was completely mesmerized by Adamina. Her crystal blue eyes, her golden-brown hair, the dimples on her smile; he could not turn away.

"*Snap out of it, lover boy,*" said Alon, man to man, recognizing Mathias' reaction.

"*Mina, why don't you escort these folks to the guest quarters, and you will get the report of what's happening in a moment,*" said Naomi.

With everything transferred into Adamina's mind with an AI meeting note-taker, Adamina escorted the Champions to a room. "*This map translates to 'Fragment A hidden in Netherheim Vapsylvania castle, Fragment B in Techtopia Technological Museum, and Fragment C in EQ*
Haven library. I can translate the coordinates for you as well; I'll just need a map of these locations."

The Codeman asked MAHIA to print a map and handed a holographic map to Adamina. Adamina charted the locations.

"*So… This mission of yours needs any recruits?*" asked Adamina.

"*We could use all the help we could get,*" said Codeman, getting excited.

"*Do you have any powers?*" asked Neveah. Adamina shook her head, "Here, we don't rely on supernatural powers."

"*Even if it's needed, don't worry. I'm sure Codeman could teach you some,*" said Alon, smirking at Mathias, making him blush.

"*I'm assuming you're Mathias… I'm sorry, I never got your names,*" Adamina said to Mathias. Mathias gave a nervous nod. As the team introduced themselves, Trophy started aggressively barking towards the balcony of the room. "*What's wrong, boy, had too much bacon?*" asked Robbie, as he ran towards Trophy to check on him; his eyes lit up, and his shutters widened. "*Holy bacons,*" a huge boom was heard. The Champions went to the balcony to take a look and saw a gigantic aircraft approaching them.

"*It was following us,*" said Codeman as he looked at the sky.

****

CHAPTER 11
FRAGMENT A

With the ships coming closer, the city faced an imminent threat. A guard barged into the room, reporting the oncoming threat of an army of skeletal creatures and huge aerial vessels. The city was under evacuation.

"Does this mean Grim has the rest of the fragments?" asked Amy.

"I guess there's only one way to find out," said the Codeman. He charged towards the balcony and jumped off the 47-story building. The Champions followed, leaving Adamina and the guard alone.

"Ma'am, I suggest you follow me to the evacuation site," said the guard. Adamina nodded and followed him out. As the Champions were free-falling meters above the ground, they deployed their jet boots. Codeman gave out hand gestures, telling them to split into groups. Amy, Aria, and Neveah were tasked with evacuation, Alon and Zeena took charge of fighting the skeletons, while Robbie, Trophy, and Codeman headed for the ship to gather intel.

On the surface, Amy and Aria provided a safe route for the evacuees to exit the area. They met up with Naomi, who assured her people that they were safe and that a nearby city had agreed to refuge them. Aria scanned the area to ensure no more people were left on the battleground, using a scanner with heartbeat detectors to find living survivors. The scanner sent out signals, and the returned data suggested the area was now clear.

"There's one person left," pointed out Amy. The system had received data that there was a person still under somewhere out there, and they were heading towards the ship at a fast speed.

"You stay with the evacuees; I'll go find them," replied Aria. More skeletons approached by the minute. While Alon struggled to fight the regenerative capabilities of the skeletons, Zeena, with her new powers, was able to disintegrate the skeletons into a state where they could no longer reassemble.

Alon noticed Aria running towards the ships. *"Why is she heading towards the ships?"* asked Alon.

Zeena turned her attention to Alon. *"I don't know. Go check out what's happening,"* she replied, turning back to the hordes of skeletons.

"You sure you can handle the skeletons on your own?" asked Alon.

"What do you think?" Zeena smirked.

Alon rolled his eyes and caught up with Aria. "There's someone left in there," said Aria, noticing Alon catching up to her. Meanwhile, Mathias, Trophy, and Robbie were getting higher and higher to the point they were losing their breaths.

"Deploy the space helmets," said the Codeman to MAHIA. The trio had finally gotten to the ship. They could see the curvature of the planet from how high up they were.

"I don't recommend looking down," said Mathias.

"Now, why would you say that? That just makes me want to look," replied Robbie.

Mathias deployed a pixy-sized MAHIA robot to orbit the ship and look for any entrance, receiving a 3D hologram of the ship. *"There, that's our way in,"* said the Codeman, pointing to a gate from where more skeletons were diving into the planet.

"That's great and all... But the jet boots' energy cells are depleted," said Robbie, introducing a technological twist to their predicament.

"Dammit, why do we always run out of energy at the worst times?" said Mathias.

"We could climb our way there with the suction grabbers," said Robbie.

"This ship is massive; it could take us hours to climb there," replied Mathias.

"*Or this,*" said Trophy as he tried digging into the cold metal surface of the ship.

"*You mean to dig into it?*" said Mathias.

Trophy nodded. "*This is ridiculous, the pup's gone mad. Who knows where we'll end up or even if we will get in,*" said Robbie.

"*What other option do we have?*" said Mathias, the Codeman. "*MAHIA, what is the thickness of the walls and how long would it take for us to get in?*" he added. MAHIA then changed her density and flew into the cold metal surface of the ship, drilling a hole into it. Moments later, she reported the ship to be five meters thick.

"*Y'know, that could work,*" said Mathias.

"*I still don't think this is a good idea,*" said Robbie.

"*Why are you being such a hater?*" confronted Mathias. "*It worked perfectly when I cut into Satan's ship,*" he glared at his robot friend.

"*When you cut into Satan's ship was the only time it worked well, don't you remember the time you cut into the trash chute, or the time you ended up cutting for hours just to find out you dug through the wrong part,*" said Robbie.

"*Done?*" said Trophy. The two turned to see a huge hole in the ship with Trophy staring from the other side, retracting his shape-shifting metal arm which was shaped as a drill.

"How did you..." Robbie said in shock.

"Cutting through ships always works a charm," said Mathias.

They went in and began exploring the ship. With the holographic blueprint and map in hand, they navigated their way to the cockpit.

Alon and Aria were closing into the location of the survivor. *"Wait. Is that...the translator girl?"* said Alon, squinting at a girl being carried by some skeletons.

"Adamina," Aria corrected Alon. Alon propelled himself into the air, landing on the skeletons and breaking them apart. Before they could reassemble, he took Adamina's hand and ran back to the evacuation site. As he made his way to Aria, a skeleton hand reassembled and grabbed his legs, causing him to trip. Soon, the other Skeletons also reassembled and sped towards the ship with Adamina captured.

When the Codeman, Trophy, and Robbie got to the cockpit, they were surprised to see the ship on autopilot. *"Wait, so they aren't here?"* said Mathias in shock. Trophy, with his enhanced senses, detected a group marching towards them. He tugged Mathias and Robbie towards a hiding spot. The three peeked at what seemed to be a group of skeletons marching in, carrying an unconscious girl. The skeletons talked to each other in sign language due to the lack of any vocal cords, which made it easy to eavesdrop on their conversation from such a distance.

"So, Grim sent them to retrieve Adamina because he needs a guide," muttered the Codeman. He sent out a message to the team, informing them that Satan does not yet have the remaining fragments. The ship is heading towards Techtopia, And Mathias, Trophy, and Robbie will be returning there, but to have the upper hand on Grim, the rest of the team should head to Netherheim.

The message was not fully understandable for the team, but the Codeman was lead and with the intel. So, they obliged, leaving the explanation for afterward.

*****

The ship landed in an old ship hangar on the outskirts of the main city. As the Skeletons exited with their prisoner, Codeman, Trophie, and Robbie had their plan ready. Trophy rolled into a cannon ball and rammed into them, causing them to knock over like bowling pins. Mathias grabbed Adamina and ran out of there. The skeletons quickly reassembled and chased them down. One of the skeletons disassembled itself and reassembled into a bike, and the other skeletons hopped on.

"These creatures…You learn something new about them every day," said Mathias, panting as he ran faster. Trophy caught up and turned his robotic arms into wheels, his harvest spat out a rope which Robbie tied to himself and the two made a makeshift sledge. Luckily the three of them knew these streets by the

back off their hand so it was easy to maneuver around and lose the skeletons. It didn't take them very long to shake off the skeletons from their trail. The trio made their way to the Technological Museum. Adamina finally woke up to Trophy licking her.

"*Where am I?*" asked Adamina. "*You're in Techtopia, my home realm,*" said Mathias.

"*How did I end up here?*" she asked.

"*Long story, but in short, you were kidnapped by the skeletons,*" Robbie tried to tell her in short.

"*But then—listen, we don't have much time. Grim could be here any second,*" said Mathias, cutting her off.

The 4 of them headed inside the Museum. "*What are we looking for?*" asked Adamina.

"*Not quite sure, but we are hoping for something similar to what was in that note.*" Replied Mathias. Robbie scanned the surroundings and concluded that there was nothing in sight that matched the input data. However, with the help of some X-ray vision, Robbie did find a secret room under one of the attractions. Trophy sensed the skeletons approaching, so he warned the team. And before they were spotted, they entered the secret passage. The group ended up in a storage area.

"*This might be where they store the older attractions or the ones they are yet to put up,*" said Mathias.

Robbie pointed towards an odd-looking spaceship,

"There. My sensors say there is something hidden there."

The four of them walked closer. *"Mean anything to you?"* asked Mathias, eyeing Adamina.

"Well, actually, this is the very spaceship that humans departed from Earth in, in search of a new habitat to call home. We thought the recruits had died, and the mission was labelled a failure," said Adamina. Mathias was left speechless. Adamina walked closer to the ship and unlocked it with ease, as she was familiar with Earth's technologies and language.

"But why would they hide something as big as that?" said Mathias.

"Because humans are pests, a plague. They forget their pasts, abandon them. They move on to infect and infest new land, slowly eating away the whole universe, making it rot," said The Grim in its rattling eerie voice.

"Thanks, by the way, your species as much as it likes to take pride in being smart… is incredibly stupid. Stupid enough to lead me right to the Artifact." It added

"You won't get away with the Artifact Grim," said the Codeman with great composure, hiding his fear inside his leadership.

"Try stop me," said The Grim. And so, the floor was again covered with mist, and the group started falling unconscious… all except Robbie.

Robbie scurried into the rocket and rummaged through the floor. Following the scans, he located the Artifact under some rubble. He heard skeletons approaching. Thinking on his feet, he cut into the floor and fell down a trash chute. From there, he managed to find an exit and ran past Grim and the Skeletons.

"*Minions…Get him,*" said The Grim.

Robbie ran as fast as he could, making his way to Aunt Billie's place. He got onto a small space pod and flew away, refueling himself on the way. Everything happened so fast, merely in seconds, that the skeleton couldn't do anything except return to the Grim with the bad news.

"*Idiotsss,*" rasped The Grim. It clenched its skeletal hands, which caused the skeletons to whirl around It and disintegrate into dust. A new group of skeletons came by, and Grim ordered them to carry Trophy, Adamina, and Mathias to the ship. The skeletons nodded and carried the three out. Another group of skeletons came by. Grim ordered them to find Robbie. And with that, Grim vanished into thin air. Going after the 3rd fragment- The fragment that could be found in Netherheim, where the team was preparing to approach without a clue of anything.

*****

CHAPTER 12
FRAGMENT B

Back on Earth, the Champions were making preparations to leave for Netherheim.

"*Will your people be okay?*" Amy asked Naomi.

"*As I mentioned, our neighboring city is more than happy to shelter us until we recover,*" Naomi reassured Amy.

Just then, Alon and Aria came running back to the rest of the Champions. "*Mathias won't be returning with us. Adamina was abducted, and Mathias, Robbie, and Trophy were on board and are headed to rescue Adamina and the Fragment in Techtopia,*" informed Aria.

"*Well, then let's pack and leave,*" said Neveah.

The Team boarded the ship. "*Where to?*" asked Alon as he jumped onto the pilot seat. "*According to Adamina, the text translated to Netherheim Vapsylvania castle.*" Alon nodded and prepared for launch.

The Champions arrived at a Crusaders HQ located in Netherheim. They stepped out of the ship, prepared weaponry and supplies, and made their way to the castle.

"*According to this map, the castle is located 6 miles on the outskirts of the town,*" said Amy, reading a map she had grabbed from HQ. The Champions decided to walk the distance to stay discreet. As they got closer, they noticed the crowded and narrow streets and alleys of Netherheim starting to become more open and emptier.

"*I've never been to this part of town,*" said Neveah. "Yeah, me neither," said Alon. The team kept walking until they came to an intersection.

"*Where now, Amy?*" asked Zeena.

Amy looked at the Map and then back at her team. "*Guys, the map ends here,*" said Amy.

"*I think we should go left,*" said Aria.

The team had no choice but to take the guess, and so they decided to follow Aria. But to their surprise, Aria had managed to guess the exact path to the castle with ease. As the Champions exited one of the dense whispering forests, they saw a huge gothic castle cover the horizon.

"*How did you manage to lead us here with no directions...have you been here before?*" Alon asked Aria.

"*Just a gut feeling,*" Aria replied nervously.

The Champions knocked on the gigantic gate of the castle. With no reply, they tried again; after still having no response, Alon forced the gate open and saw a group of people fully armed on the other side, ready to attack.

"Wait, Wait, Wait. We come in peace," said Alon, raising his arms.

"Aria...Aria Shade, is that you?" asked one of the girls standing on the opposite side.

"Wait, you know them?" asked Zeena.

"Of course, she knows us; she grew up with us at Vapsylvania Academy," another girl said.

"Are we going to just stand here, or can we come in?" asked Aria. "Come in... Miss Prudence would be delighted to see you," said one of the guys. The Champions, with their trust in Aria, entered the castle. They followed Aria and the other group to a lounge.

The room was covered with dim, warm lights of old torches. There were green couches circling an ugly purple rug. But the room still gave off a warm and cozy feel. The Champions saw another boy preparing tea through a small door, which led to a partially open kitchen. The guy brought in some floating teacups and a teapot. The boy poured everyone a cup of tea and sat down with the rest.

"So, is anyone going to explain what the hell is happening," said Neveah. *"I'm sorry I should introduce*

you guys," said Aria. "The girl with the blue hair... Her name is Vanessa. She has the power to manipulate and control Water in all its forms, whether liquid, solid, or gas. The girl with the afro, her name is Camella, she can shapeshift into any human or animal. The guy with the leaf sprouting from his hair is Gordon. He can control plants and rocks. The one with the scar on his face is Thomas; he can talk to Ghosts and can turn into one and lastly, there's Gus. He's the one who just served you guys tea. He has wind powers, but it's basically telekinesis."

"I'm assuming you already know the Champions," said Aria, looking at the group.

"Yeah, we heard about them at the Portal opening ceremony," replied Gus.

"There's a lot I haven't told you guys about my past before EQ University...This happens to be one of those things," said Aria. "I was raised an orphan at this academy. This is where I learned all my magic," she added.

Aria went on to explain her origins: "Vapsylvania Castle was built by Empress Barbara for her adopted children; after their deaths, it was repurposed into an orphanage by the 3rd adopted child, Prudence. Prudence claimed to have a disability, one where she could not perform magic. However, she dedicated her life to teaching orphans such as herself magic and all *about it. At least that's what she has us believe,"* Aria explained.

"Not with your conspiracies again," said Camella.

Aria gave Camella a side-eye and continued explaining, *"Prudence has been manipulating the orphans in her orphanage through brainwashing. Her claiming that she is powerless is all a ruse. She just wants our sympathy, and she uses mind-control powers to gain it. She does not care about raising orphans; all she cares about is selling them, raising them into weapons."*

As Aria finished explaining, she looked back at her old peers and saw how they still didn't believe her. Gus rolled her eyes at her theory and asked for proof.

"I have been giving you guys the same statement this whole time, and it remains the same as every time. I secretly applied to EQ University to further train for magic and even expand my field by learning more about technology. When my results came in, the letter said I passed. I wanted to surprise Prudence, so I went to her that very evening. When I revealed to her my accomplishment, I thought she would be happy, but no, I was wrong. She was furious. She didn't want anyone to escape; she didn't want us to have our own lives; she didn't want us to have free will. I ran away from the room and began packing anyway. I cried; I didn't want to leave my only family, but I had also worked so hard for this. When I was about to leave, I heard Prudence approach me, so I made a shadow

clone of myself and hid in the darkness; I saw as she tried to hypnotize me, try to get me to forget I ever applied, and to convince me to stay. Her eyes turned a bright yellow with black spiraling pupils. While she was trying to hypnotize my clone, I tried warning you guys you were willing to leave with me, but she caught up. She hypnotized you guys to forget what I said and to not believe me. I barely managed to escape that day."

As Aria finished telling her story, Miss Prudence walked into the room. *"Camella, you called?"* asked Prudence.

But as she turned to Aria, she was shocked to see her. But more importantly, she was furious. It was obvious from the way she gritted her teeth. However, she swiftly composed herself so as not to blow her cover.

Aria, splendid seeing you here…And you brought friends, I see," said Prudence.

"Don't try to sweet-talk me into believing you. I know what you did," said Aria. *"And if you try to hypnotize me again, I hope you know that I trained to fight back. Your hypnosis will not affect me,"* she added.

"Is that why she asked me to try to hypnotize her?" Wondered Neveah, looking back at their training sessions.

"*Well, what brings you here?*" said Prudence, ignoring Aria's comments.

"*We are here to find an Artifact, a fragment of a key of sorts,*" said Alon, attempting to bring the focus back on the mission. But Prudence instantly denied the presence of such a key. However, Neveah could read her mind. She could tell Prudence was lying. Aria ignored Prudence's response and again used her scanner to find hidden areas in the castle.

"*Anyway, Aria is back after such a long time, and you are her friends. You are our guests,*" Prudence said with a highly fake sweet smile. "*Gus, Camille, show them the guest room.*" She added, turning towards her wards.

The Champions looked at each other. Just as they were parting toward their separate rooms, Aria whispered just two words: "*in fifteen.*"

After exactly 15 minutes of entering their rooms and assuring Gus and Camille that they would manage, the group regathered in Aria's room.

OK, she's lying," Neveah declared as everyone huddled to make a plan and look around to find their target. They left the room and followed Aria around as she snooped, looking for the fragment. "*How did you live here for that long?*" Amy said as she looked around in the dark and dusty hallways of the castle.

"*I bet it's haunted as well,*" said Alon, moving a bit closer to Aria. Aria rolled her eyes and elbowed him sharp in the rib. "*Guys, I think I found something,*" said Aria. Alon rubbed his sore spot and followed her, boring a hole in the back of her head only with his glare. "*There…Again…It blinked.*"

"*I mean if you really say so. That guy Thomas did say this place has ghosts. Personally, I don't bel-*" said Amy before she was cut off by rustling and footsteps behind the painting. The two quickly removed the painting, which revealed two eyeholes.

"*Well, at least it ain't ghosts,*" said Amy.

"But then, who was watching us?" asked Gus.

Meanwhile, Zeena and Camella ended up at a library. "*Ugh! Its always libraries,*" groaned Zeena. Zeena used a summoning spell with reference to the screenshot of the drawing of the artifact on the map. "*Let's hope we're in proximity. If the fragment is in here, it will be in my hands on my summon,*" said Zeena. The two heard shuffling among shelves before one of the shelves collapsed, and the fragment flew right into Zeena's hands.

"*Girl, you are cracked,*" said Camella in amazement.

Thanks, now let's inform the others and head back." Aria and Thomas ended up in a dungeon.

"I'm getting a lot of paranormal activity in this room," said Thomas. The two started checking cells. As Aria opened her 6th cell, she suddenly got a huge headache. Thomas grabbed onto her before she fell unconscious. Aria woke back up to see Amy using some healing magic on her.

"You alright?" asked Amy. *"Yeah, I'm fine, thanks for coming,"* said Aria as she got back up.

"You released a ghost," said Thomas. *"I...What?"* asked Aria. *"There was a spirit trapped in the cell, A very strong one at that. When you opened the Gate, the rush of the paranormal activity knocked you out. I believe the spirit might have attached itself to you in order to leach off your powers and to escape."* He explained.

"Well, then, get it off, you're the expert," said Neveah.

"I will...But first, we need answers. The reason we even came here was to find out the secrets of this castle," said Thomas.

"Ugh! We don't have time for this. I'll just destroy the spirit," said Zeena as she clenched her fists, manifesting some of her cured corrupted magic.

"I'd be careful with that. If you destroy the attached spirit, you might strip her powers or kill her off," said Thomas.

"Fine, I'll do your stupid ritual," said Aria.

"*Great,*" said Thomas. Thomas drew up a ritual circle on the floor. "*No one enters the circle except Aria,*" said Thomas. Everyone exited the circumference except Aria and Thomas. Thomas held Aria's hand and chanted a spell. The Spirit that was attached to Aria flew off and crashed on the circumference. The circle had made a barrier that didn't allow the spirit to leave or attach itself to anyone else.

"*Who are you?*" asked Thomas.

"*I'm Charles. The 2nd adoptive son of Barbara*" rasped the spirit. "*Ok, Charles, I have the power to set you free, and I am willing to release you if you answer my questions…is that clear?*" said Thomas. Charles gave an uncomfortable nod.

"*Why are you here?*" asked Thomas.

"*Prudence…She trapped me here…Starved me to death,*" hissed Charles. The orphans took a step back in shock. Aria gave them a 'Told you so' look. Charles went on to explain, "*Prudence had us all fooled. She claimed she didn't have any powers, but she did. The power to manipulate. She used it for years on Mom. That's how she got Empress Barbara to adopt, and that's how she ruled over Netherheim for years. She used us all as puppets. When Mom died, she lost her power. An adoptive child wasn't allowed to take over the throne. So, she went after the next best thing… The castle. She trapped me and the 1st son, Philip, down here so that she would get ownership of the castle. She wanted to start this program. She wanted to train children to be powerful magicians, and then she would brainwash them and sell them as slaves"*

"We've heard enough…You're free to go," said Thomas. He made a hole in the barrier and supplied Charles with enough energy to be able to roam freely in the walls of the castle.

"I guess you were right," said Camella, looking at Aria.

"Well, then, where do we go from here?" asked Gus.

"You could always join us at the Champions," said Amy.

Aria, Zeena, and Neveah looked at each other awkwardly. It was not their decision to make in the absence of the Codeman. But Thomas said,

"Thanks…We'll consider it."

"Where's Alon and Gordon by the way?" asked Aria. *"Guys… we have a problem at the entrance,"* radioed Alon. The group rushed to the main room and saw an Army of Skeletons with Grim right behind them.

"Grab this and run," said Zeena to Aria, handing her the artifact.

Aria grabbed the artifact and disappeared into the darkness. Zeena immediately started disintegrating the skeletons while the others escaped and covered Aria. The Champions and the orphans got back to the ship and began boarding. Neveah stopped and called out to the others, "*I will stay to help out Zeena.*"

When Neveah returned, she noticed that there was only Zeena and Grim left on the battlefield. Grim was using its usual tactic of making Its victims fall unconscious, but Zeena was resisting with all she had. In desperation, Neveah used her powers and tried to get into Grim's head but it wouldn't work. Grim was immune to telepaths. Grim, taking notice of Neveah, started targeting her as well until the two were too tired to fight back. Grim then waved his arms, teleporting the two back to his ship and then disappearing himself.

On the other side, just when the Champions were about to take off, they noticed a space pod approaching them. It was Robbie. The Champions opened the door for Robbie to enter and were shocked to see him alone. Robbie was shocked to see the new people on board. Everyone was shocked and puzzled and looking at each other, waiting for the other to explain. Where was the Codeman.

****

CHAPTER 13
GATEWAY TO GODS

"Get away, there's more of them," said Gordon, aiming his hands towards Robbie to mold his body's metals.

"No...Wait, that's..." Amy said, but it was too late. Robbie's exhausted body started collapsing into itself and crumpled like aluminum foil.

"He was on our side," said Amy, looking down at the crumpled body.

Aria walked towards Robbie and grabbed his CPU and Memory drive. *"Don't worry, it's nothing that can't be fixed,"* said Aria as she pulled out more parts to salvage. As she picked apart the scraps, she saw a strange-shaped object.

"Is this..." Suddenly, the object and the Fragment the Champions had recovered started glowing and ringing.

"That's the other fragment," said Aria. Alon, who was holding the Fragment they had just recovered, walked closer. The two fragments pulled towards each other and conjoined.

"That's two fragments down; all we need is the last one, which is with Grim," said Amy.

"How do we plan on finding that fragment…Not to mention, we also need to get Zeena and Neveah back," said Alon. *"We'll figure that out soon; for now, let's focus on getting Robbie back so we can find out what happened to Mathias and Trophy,"* said Aria. *"In the meantime, toss that fragment in storage and give our guests a tour,"* she added. Alon nodded and the rest followed him into the hallways of the ship.

Mathias woke up to Trophy tugging on his pants. *"Where…Where are we?"* said Mathias, rubbing his eyes. He stood up and looked around, noticing Adamina unconscious on a stone bed. *"Can't believe we're captured again; I hope Robbie's safely made it to the others,"* said the

Codeman, patting Trophy. Mathias noticed that Trophy was having trouble breathing; he looked down at Trophy to see his cyborg parts, which helped him live, were taken away from him. He grew worried, *"MAHIA, give us a scan of the area,"* ordered Mathias. But to his astonishment, there was no reply. He then repeated, *"MAHIA, give us a scan of the area."* There was still no response. The Codeman looked back on himself and noticed he was no longer in his suit of armor. Growing increasingly worried about their current situation, he aggressively banged on the door of the dungeon. After minutes of banging, he heard footsteps approaching from the outside.

"*1...2...3...4...5. 5 people,*" he said as he listened to the footsteps. Mathias looked around the room for something he could use to immediately strike whoever was approaching. He could see a few exposed pipes and started investigating. The footsteps got louder and louder; Mathias had to hurry his plan if he hoped to succeed in escaping. He listened to the sound of the footsteps echo throughout the hallway. Mathias checked if the pipes had any running water or gas until he found one that was empty. He kicked the pipe until he felt like his foot would drop off his leg. The pipe finally fell down. The Codeman told Trophy to get back and protect Adamina. As he got closer to the door, Trophy barked. Mathias looked back at Trophy, lowering his guard; just then, the gate of the dungeon opened, and three guards threw Zeena and Neveah into the dungeon.

"*You?*" said Zeena, who was being untied by the guards. The guards untied both Zeena and Neveah and again locked and sealed the door.

"*You guys got captured...Where's Robbie?*" asked Neveah.

"*He escaped with the fragment to meet up with you guys, but it seems like you all got captured as well,*" said the Codeman in frustration.

"*Nah, it's just us here. The others made it out safe with the artifact,*" said Zeena.

"*Hope Robbie made it to them,*" said Neveah.

"We have to get out of here," said Mathias.

"Don't worry. The Champions will probably come get us soon," said Zeena.

"No...no you don't get it. We can't wait around for that, Trophy...He's dying... Without his bionic parts, he could die any minute," said Mathias in desperation. *"Besides, I told Robbie not to return,"* he added.

"What about magic..." said Neveah.

The Codeman tried to cast a short-range teleportation spell, but his spells wouldn't work. He tried more and more spells, but yet again he had no luck. Neveah tried a few of her own spells but none worked.

"I hate magic barriers," said Neveah. Zeena even tried her cured magic but even that had no effect against whatever barrier Grim had cast. The Codeman again punched the door in frustration. All the noise woke Adamina up.

"What's going on?" she asked.

"We're trapped in a cell, and our friend's dying," said Zeena in a monotone voice.

Adamina walked closer to investigate the door. *"Is it completely shut and sealed?"* she asked.

"Seems like it," said the Codeman.

Adamina knocked around the door and tried to find weak spots in the old-fashioned Earthly way. *"Here, strike here,"* she said.

The Codeman used his other foot to kick the door as hard as he could; a couple of kicks later, the door was dented, but Mathias could no longer kick.

Zeena swapped in for Mathias and continued kicking until a part of the door fell, revealing wires and machinery. Adamina took a close look and after a couple of moments, rewired the door into opening. *"There… Done,"* she said. Mathias, Zeena, and Neveah looked at each other and the Codeman thanked her.

The four of them walked out while Neveah carried Trophy, who was now unable to even walk. Mathias who was also limping due to the damage his foot had taken, walked with the support of Zeena and Adamina. Trophy barked and warned the others of approaching skeletons. Zeena let go of Mathias, leaving him to Adamina, and prepared to face the skeletons. She began fighting with everything she'd got but they just kept rearranging themselves. Zeena now again kicked the skeletons, and this time, before they could rearrange, she snatched pieces of the bone and scattered them around while keeping the spine, one of the vital parts of the skeleton, as a weapon. Adamina also grabbed one of the skeleton's fingers so they could easily access the doors through biometrics.

At the Ship, Aria looked for parts to build Robbie's robotic parts but the ship was low on resources. She scoured the whole lab for any parts, but there was simply not enough. As she looked around, she noticed a spy drone on her desk. That gave her an idea. She rewired the drone and made it so that it was able to have Robby's memories, personality and mind. She also plugged in a speaker on the drone, allowing it to speak.

"*Seriously…This is my body?*" asked Robby sarcastically.

"*Oh, piss off. It's only temporary, and be thankful I gave you one in the first place,*" said Aria.

The two exited the room to meet up with the others. They arrived in the cockpit, where everyone else waited for them. "*You guys had something to say…*" asked Alon.

"*What's happening?*" said Aria.

"*We want to join the Champions if that is ok with you,*" said Thomas. The others nodded.

"*You guys just achieved freedom; you are no longer just weapons, yet you want to do exactly that?*" Aria asked.

"*This time, it's different. Prudence may have tried to manipulate us and try to make our weapons against our will, but this is our choice; magic has been our passion, and we always wanted to use magic combatively for the greater good,*" said Gus. Aria was about to explain that the decision was up to the others, particularly the Codeman. However, the moment was soon disturbed by a loud explosion.

Amy turned back, *"It's Grim,"* she exclaimed.

"He's probably here for the fragments," said Aria.

"Do we engage in battle?" asked Vanessa.

"No, even with all the manpower we have, we are no match for Grim; we must escape at once. But we must leave undetected." Aria said. *"There is a teleportation portal located in the lower parts of the ship next to cargo. It is out of order, but I can try to get it working,"* she added.

Right as the Champions went ahead to make their way out, sirens went off. "Intruder alert," repeated one of the speakers.

"Dangit, it's here," said Alon.

Aria used her spell to hide themselves, and the team sneaked their way to the portal. Luckily, it seemed Grim had entered from the west wing while the team needed to go to the east wing. Aria immediately got to work on the portal.

"Where's the fragment, by the way?" asked Amy. The Champions went pale; the fragment was still back in storage, which was right where Grim had infiltrated the ship.

"Don't worry, I'll go get it," said Thomas. *"It's too dangerous to go alone,"* said Amy. But then Thomas just collapsed.

"Too late," Aria rolled her eyes.

"What just happened?" asked Alon.

"Thomas's spirit or soul can leave his body any time he wishes; essentially, he just leaves his body as a ghost; this allows him to phase through walls, turn invisible, and possess anyone. He can also immediately get sucked back into his body no matter where he is," Aria explained.

Thomas made his way to the west wing of the ship, going past many skeletons undetected. Luckily, he remembered the tour Alon had given him and made his way to storage with fair ease. He entered the storage facility and saw Grim scouring through drawers and cabinets in search of the fragment. Thomas could feel the negative aura oozing out of Grim, but due to him being undetected and having no body holding him back, he was able to stay conscious around Grim. Thomas put his hand through the cabinet and held the Fragment; he materialized his hand, which meant now he was making physical contact with the Fragment.

"Some villain," he rolled his eyes and returned to his body in an instant. Thomas woke back up in his body with the Fragment in his arm.

"And...Done," said Aria.

She booted up the portal, and the Champions made their way back to EQ Haven. Thomas again left his body, which Aria carried back through the portal.

*****

As soon as everybody had left, Thomas destroyed the portal so they couldn't be followed back. He started roaming the ship, sabotaging anything he could, and even snuck onto Grim's ship, which would lead him right to Zeena, Mathias, Neveah, Adamina, and Trophy.

*****

After wandering for what felt like hours, Mathias started realizing the structure of the ship. He started memorizing the paths, mapping out routes and even tried to remember the blueprint MAHIA had made the last time he was here. Trophy had now lost almost all of his energy and was unconscious. Soon, the four reached storage. The room was guarded by more skeletons. The Codeman instantly felt like he was close enough in proximity to summon his armor.

"MAHIA AUTOPILOT!" he shouted. Suddenly, the disassembled armor that was sat atop an altar, assembled on its own, knocked out all the skeletons and threw them out a trash chute.

"Almost did my job better than me. AI really is coming for our jobs," said the Codeman as his armor approached him. He boarded the armor, collected Trophy's parts, and assembled them on him. *"Please still be with us, buddy,"* he said as he activated the parts and read the vitals. Mathias's helmet's visor started fogging up from tears of happiness. Trophy was alive; however, he was still in a coma.

"We need to get back fast; Amy can probably fix him," he said. The three nodded. Mathias tried reaching out to the Champions but it seemed the signals were jammed. The four of them heard a loud thud of a smaller ship connecting back to the main mothership. They sneaked forward to investigate and saw Grim and Satan arguing.

"You let them escape? You fool," hissed Grim. *"You had one job,"* it added. *Now, find them,"* said Grim as it walked away. Satan rolled his gruesome eyes and assigned every skeleton to be on the lookout.

Neveah felt a gentle touch on her back. She turned around and went pale. *"Is that a...a ghost?"* she gasped. The Codeman took a stance to fight.

"Thomas? Don't worry, he's with us," said Zeena.

"It's a long story," added Neveah, looking at the Codeman's confused face.

"Have you found the fragment?" asked Thomas. The three shook their head.

"Well, then, let's get to it. We need to find it before it finds us," said the Codeman.

The four of them sneaked around the ship. With MAHIA back, it was easy for Mathias to do a scan and locate the fragment. The four followed a map that MAHIA had created and ended up at a vault. Adamina took a step forward and cracked the vault with ease.

"How do you know so much about vaults and doors?" asked the Codeman.

"Oh please, me and my grandad have been going to escape rooms since I was 6. I love puzzles like these," she replied as she giggled. Mathias felt his heart race at the sound of her tinkling giggle, but the moment was quickly interrupted.

"Going somewhere?" said a familiar voice, breaking the excitement. It was Satan.

"I'll deal with him. Go get the fragment," said the Codeman. He booted up his flamethrower and started shooting it through the narrow vault door. Satan, who was unable to get inside, sent his skeletons to fight. As soon as everyone was inside the Vault, the Codeman quickly closed the vault on himself. Adamina rushed towards the vault to change the locks.

"Watch out. It's hot," said the Codeman. He handed her a gauntlet, allowing her to safely lock the door.

"That should hold him for a bit," said the Codeman.

"Found it," yelled Zeena.

"Great, now, everyone grab onto me," said Thomas. Knowing it was either trusting this stranger or facing Satan, an army of skeletons, and quite possibly even Grim, the Codeman agreed. Thomas read out a spell and so he was now back in his body, with everyone still holding on.

"Wow, what a portal," Zeena giggled, earning a burning glare from the Codeman. *'Yeah right, you can be all eyes for a stranger girl and we can't even joke with a guy who's not that much stranger to us.'* Zeena thought gritting her teeth but didn't say anything.

"You guys. You're safe," said Amy. The Codeman picked up Trophy and handed him to Amy, *"Save him...please,"* he said. Amy nodded.

"I'll grab that off of you," said Aria, referring to the fragment in Zeena's arms.

"It won't be a while till Grim is back for these. We have to move and fast," said the Codeman.

"Let's head back to HQ and cook up a plan then, regather our thoughts and gear up again," said Alon.

The rest agreed. Mathias stared into the sky and thanked whatever deity looked after them, for what seemed like an impossible mission was now seeming very hopeful. Time and time again, they had escaped from the clutches of death; if only they continued down this path, they might just beat Grim and Satan. They had now the fragments, a Gateway to Gods.

****

CHAPTER 14
SEEKING ASSISTANCE

The Champions gathered around the meeting room one last time. The Codeman placed the now-joined fragments on the table.

"This key is what Grim has been after. It's essentially a Gateway to Gods," he said, staring down at the key. *"What if we use it instead and ask the gods for assistance? We can't beat Grim on our own, but surely the Gods can... They are the ones who created it in the first place,"* Codeman added, looking at everyone for their opinion.

"But what if they refuse?" asked Alon.

"It's not like we have much of a choice anyway," said Aria.

"Wouldn't we be helping Grim by opening the Gate? Isn't it best to stash the fragments and hide them?" said Amy.

"That solution has already been tested, and Grim found them all. We can't run from it anymore. We have to fight and defeat it once and for all," Codeman responded, looking around. His eyes met Adamina's

over the table. The admiration in her eyes filled Mathias with a beautiful warmth, like honey melting from his heart to the core of his being.

Mathias knew his feelings for the new blue-eyed girl. He was also aware that he had been attracted to so many girls in the past, including Zeena, but he was not a boy anymore. And the man in him was certain about his heart's desire. He had never felt like this before.

Nevaeh's voice brought him back to the matter at hand, and he directed his attention to what she was saying.

"Well, then, let's take a vote," said Nevaeh.

"All in favor, raise your hands," said Zeena. Although hesitant at first, the Champions trusted the judgment and decision of their leader and agreed to his plan.

After the meeting, Amy and Mathias walked back to the Medbay to check on Trophy. Amy had just installed the cyborg parts on Trophy and had given him some medicine to assist in breathing and readjusting his body. Trophy was breathing normally, but he was still not fully awake.

"Stay here with him," said Codeman to Amy. Amy nodded. Mathias left the room and went to the armory, where the other Champions were gearing up.

The Champions gathered all the materials and weaponry they needed and supplied their newer members with armor and weaponry.

"Let's get out of here before Grim gets on our tail," said the Codeman as he walked to the Garage. He put his palms on the scanner, unlocking the Garage door.

"You might wanna…" said Alon before he was interrupted by Codeman's confused voice.

"Where's the ship?" he said.

"Grim hijacked us. We lost the ship," said Aria.

"Well then, how do we get to the gate?" Codeman asked. Aria eyed one of their older, janky vessels.

"Oh, you have got to be kidding; no way are we going in that; it's on the verge of turning into dust," said Alon, and Codeman nodded in agreement.

"I recall someone talking about limited options a while back," said Robbie, hovering over Mathias in his drone body. Mathias rolled his eyes, and the Champions began squeezing into the ship.

"We really need to build a portal on Earth as well. Travel by ship can be annoying," said Nevaeh.

Codeman hopped into the pilot seat and booted up the ship. *"According to the map, the gate should be in the temple where we ended up on Earth the very first time,"* said Aria. *"Well, then, that's where we're headed."*

The janky old ship zoomed through, exiting EQ Haven. Due to the ship being a very old, worn-down model, it was not meant for space travel. However, the Champions were too late to realize that.

"I don't mean to alarm you guys...but the ship is losing some parts," said Robbie, hovering around the windows as he watched plates and sheets of metal fly away from the ship, disappearing in the empty void of space. The odds of the Champions reaching Earth seemed challenging, but it was too late to turn back, and they had to push through. By the time the team had arrived at the solar system, they had lost most of the ship.

"We're almost there," said Adamina, recognizing the planets around her star system. However, all went downhill when the Champions reached the Asteroid belt, a ring or orbit of asteroids between the planets Mars and Jupiter. The Kuiper belt, the belt on the outskirts of the solar system, had already damaged the ship enough, but the asteroids in the asteroid belt had struck the final blow.

"Guys...I regret to inform you, but we have lost our engine," said Robbie.

"We what?" said Alon, slowly turning towards Robbie. Suddenly, the ship started showing all kinds of warnings and alarms. The Champions had no choice but to crash into the nearest planet, Mars. They

dodged and weaved through the asteroids and exited The Asteroid belt with what remained of the ship.

"Prepare for a crash landing," yelled the Codeman. The ship closed in on the rusty red planet and was slowly pulled in by its gravity. As it got closer, the ship started accelerating until it finally fell onto the planet, breaking into pieces.

"Everyone OK?" asked Mathias. He looked around, watching as everyone showed a thumbs up.

"Honestly, it's a miracle we made it this far anyway," said Aria.

While the rest of the Champions were off to visit the Gods, Amy tried her best to bring Trophy back to full health. Amy went back to her room to look for some medicine and noticed a spell book on her nightstand. She remembered how she wanted to learn magic like the rest of her peers but never got the time with all that was happening around her. It was still crazy to her that she had been dragged into all of this and how drastically her life had changed ever since she joined the team. Before the Champions took her in, she was a medical student who had just become a doctor; she had a fulfilled life and plenty of time for herself, but she always had this goal of helping as many people as she could. But when Satan arrived in Techtopia, her calm and normal life was taken away. She watched as she and her peers were persecuted and tortured. That made her realize she didn't want anyone else to ever go through that. Amy snapped back to the present. She thought about using magic to heal Trophy, but there was self-doubt that she might mess the spell up. Nevertheless, she convinced herself to take that step, to help more people. She snatched the spell book off her nightstand and rushed back to the Medbay. She held up her hand, closed her eyes, and recited the spell. As she finished reciting, she slowly opened her eyes, lowered her arms, and saw Trophy sitting up, panting. "*I did it*," whispered Amy. "*I did it!*" she said again, this time louder, celebrating. She picked Trophy up and spun him around in happiness. However, that happiness didn't last long. Amy heard a huge thud from outside HQ; she peeked outside and saw that Grim was here. "*I have to stall them as long as I can; you go run, you have just healed, and you can't run into danger again*," said Amy to Trophy. Amy put Trophy in a small space pod and sent him to Earth, hoping it'd be safer than EQ Haven. Amy exited the building; she stood facing thousands of skeletons and once again opened the spell book, in search of offensive spells. She walked towards the skeletons and recited another spell, this time the spell sucked away the health and buffs of the skeletons and added them onto her, meaning now Amy's skeleton was as strong and durable as her opponents. Whereas, the skeletons' bones seemed weaker. Amy dashed towards the skeletons, and with her knowledge of the human body, she disassembled each skeleton one by one till she neared the ship. Amy had easily knocked out most of the skeletons, but she was bound to get overpowered at some point. She was severely outnumbered but still pushed through. The battle, despite the numbers, was on Amy's side but as soon as the sky went dark, and the mist covered the ground, it was over for Amy. Grim walked out of the ship, staring Amy down. *"Impressive, you knocked out most of the army; tell you what, out of respect, if you tell me where the artifact is...I might just let you live,"* said The Grim.

"You must be another kind of delusional if you believe I'll ever tell you where it is," said Amy.

"So be it," said Grim. Amy knew she couldn't take Grim on, so she whispered a spell in her last moments. It was a spell that would keep healing her as long as she didn't take a critical hit, a hit that could kill her in one strike. Grim was known to use spells that would slowly poison you and make you wither away, so Amy took the gamble and hoped the poison would be canceled out by the spell. Grim used his spell and walked away. He knew the other Champions had left, and luckily, they had left a trail that would lead him right to them, through parts of their ship floating throughout space. The fleet of ships quickly left the landscape as Grim and its army went away to retrieve the artifact from the Champions. Amy quickly got up and shook off the last bit of poison. She went to the garage, looking for any other ship or pod, but she was out of luck. She had no other choice but to use a portal and try to retrieve a

new ship. She recalled that her parents had a ship, but she hadn't seen the ship or her parents ever since the incident with Satan. She limped her way toward the portal and entered the coordinates for the nearest portal to her home. As she stepped inside the gate, memories flooded back of her home. Despite all Techtopia had been through since her last visit, nothing had changed there. She went inside her house and yelled for her parents.

Her home was a symphony of AI convenience. Stepping through the doorway, which hissed open in recognition of your retinal scan, you were greeted by a cheerful *"Welcome home!"* from the central AI, nestled discreetly in the ceiling. She entered her once aromatic kitchen, which now looked the same as before but abandoned and devoid of the aroma of her mom's baking. Sleek surfaces responded to your hand gestures. The refrigerator, a seemingly blank panel on the wall, pulsed with recognition as you approached, displaying its holographic contents.

A small, spherical bot was stationed, programmed to diligently carry laundry to a concealed cleaning unit. Overhead, light levels adjusted automatically, mimicking the soft rise and fall of the sun outside.

But there was no answer. She tried again and again there was no answer. Her neighbors heard her yell, and a girl approached her. *"Amy…is that you?"* said the girl. Amy turned around to face the girl. *"Leah?"* she asked, recognizing the girl's voice. The girl was Amy's childhood best friend and they even attended med school together. *"Leah, have you seen my parents?"* said Amy in a panic.

"It's a bit late for that…Where have you been?" said Leah.

"Too late…What is that supposed to mean?" said Amy, starting to tear up as she realized what Leah was talking about.

"Your parents were discovered dead, among many others in Techtopia after the massacre," Leah replied ruefully.

"Take me to them," said Amy. Leah nodded. The two took a walk around their neighborhood and to the local graveyard. Amy looked down at her parents' grave and burst into tears.

Meanwhile, the Champions had been walking around the rusty red planet for hours now. Adamina informed the Champions that there was a city built further north. It seemed she had a lot of knowledge about Mars and other planets of her solar system, hence was selected as their guide. *"We're here!"* yelled Alon.

The others caught up to him and saw the city popping up on the horizon. The city was covered with a bubble dome around it which captured an atmosphere allowing the humans to breathe. Inside the dome were large skyscrapers similar to Techtopia's older buildings. And although the skies were dark, the city was covered in bright neon lights. The Champions got inside the city and started looking for anyone who could get them back to earth. Adamina opened her phone and started looking for transport services.

"*What's with the box?*" asked the Codeman.

"*I'm sorry it's not in a fancy hologram for you,*" said Adamina.

Mathias went red and said, "*No, no I didn't mean it like that, I'm just curious about Earth's technology.*"

"*Don't worry, you're good. I was only just joking,*" said Adamina.

"*Guys, this man can drive us home,*" said Alon, pointing towards some older-looking guy giving a thumbs up.

"*Can you even understand him?*" asked Aria. Alon shook his head. Adamina walked past Mathias and talked to the guy and agreed on a price. The old man took the Champions out of the city to an old ship hangar and launch site and told the Champions to hop on board.

"*All this seems a bit shady,*" said Zeena.

"*Not to mention what happened the last time we followed an old person blindly,*" said Nevaeh, referencing the descendant of Zerlin.

"Come on, guys! We're Champions! We can handle a few old dudes and their rusty spaceships. Besides," he winked, *"we have magic, and they don't."*

His words, though intended to lighten the mood, did little to quell the disquiet settling in their stomachs.

The old man led them into a spaceship, a homey and comfortable setting.

Despite the comfy, cozy interior of the craft, exhaustion hung heavy in the air but beneath it.

Beneath all this, a current of nervous energy crackled between Adamina and Mathias. Their gazes kept locking. Suddenly, Adamina saw a kitchen area peeking out of the corner of a corridor. She shuffled to the corner, feeling Mathia's eyes following her. She opened some cabinets and shouted back. "*I found the tea supply. Let me treat you guys with the ancient brew. It's tea, and it will freshen up your mind.*"

Mathias appeared beside her, a ghost of a smile playing on his lips. *"I'd love to help,"* he offered, his voice husky.

As they rummaged through the dusty canisters, their fingers brushed and a jolt of electricity shot through them both.

"*Where is the famous tea?*" Alon entered, oblivious of the current crackling the atmosphere. Mathias swore under his breath and smoothed down his hair.

And left the kitchen, dragging Alon behind him to give Adamina some time to compose and arrange herself.

A few minutes later, a composed Adamina entered the main deck with some steaming cups. The old man flew them to a country on Earth called America. *"This place seems very different from where we first arrived,"* said Alon.

"Earth used to have multiple countries, but in an event called Pangea, the continents were joined. However, there were still factions who didn't wanna join, so they just established different countries as a combination of all countries. It's a complex matter, and it lasted a couple of decades, so it's hard to sum up; the current countries are Africa, America, the European Union, Asia, and India," explained Adamina. *"No worries, though; there is a bullet train to Asia just a few meters ahead,"* she added.

"More walking?" wailed Alon. The Champions ignored him and got to the station.

"What's with the traffic," said Mathias.

"We're still working on public transport, even after centuries. It's Earth culture at this point," Adamina answered, trying to shake away the effect of Mathias' closeness in the kitchen, acting composed.

****

It took them some time to reach Asia and the destined temple. They experienced riding an Earth train for the very first time, along with various things that they had only heard in folklore.

As they got there, they saw an asteroid approaching them. The only thing was that it wasn't an asteroid; it was Trophy.

"Trophy, you're awake. How are you, my boy?" said the Codeman, vehemently patting his head and pinching his cheeks. *"Wait, why are you here though,"* he added.

Trophy barked, and the AATD said, *"Grim."* The Champions knew that Amy had sent Trophy to alert them that Grim was coming, and they had to act fast. Codeman rummaged through his bag and pulled out the key made up of all three fragments. The Champions walked inside, and to the portal, they had found a year ago. They placed the key inside a mold and recited a spell that was written on the map. As they finished, the portal lit up, a crack or rift appeared. The rift expanded and revealed the realm of gods.

"Here goes nothing," said Mathias as he and the rest walked inside.

****

CHAPTER 15
GOODBYE GRIM

The Champions walked in and were met with a huge golden gate between them and heaven, the home of the Gods. There was a huge door knocker that Mathias, the Codeman, used. As he knocked three times, the God of Gateways and Secrets appeared.

"What is it that you seek?" said the God. Codeman explained their situation in the shortest time and with the fewest words possible. However, the mention of Satan and Grim jolted the God. Doubtlessly, their situation was worthy of the Supreme God's time. With a clap of his hands, the Champions found themselves in a courtroom with all the Gods sitting around in a circle and the Supreme God on a throne, waiting to hear the Champions' plea.

The court had multiple statues and banners showing the Supreme God being worshipped by other gods, his angels, and humans. The God of Secrets whispered something to the Supreme God and moved back, bowing his head. The Champions looked at each other, finding such submission overwhelming in their age and time.

"So, you are here to discuss Grim? Ah, yes, the angel who, though he knew better, was kicked off right alongside Satan; that imbecile is probably after the Fountain of Godhood; we aren't too concerned. To reach the fountain, you must complete many trials that he could not even come near to completing," said the Supreme God. However, Codeman was not reassured. Grim was one of their most powerful opponents, and he still harbored doubts despite the Gods treating the situation so lightly.

The Champions were getting ready to leave after being humiliated by the Gods, deciding it best to follow up with the plan of destroying the portal and the key, scattering them across the realms. As they were about to step out the door, the God of the Cosmos sensed something. He jumped out of his throne onto the ground and rushed out the gate, passing the Champions. The Champions followed the God who was looking up at the sky. They also followed his gaze and saw a moon rising up, blocking the sun; it was a total solar eclipse.

"These aren't supposed to happen, not in our realm, not in heaven," said the God of Cosmos. The Champions heard loud sounds of thunder, and the floor of clouds beneath them started turning smoky gray with purple lightning and thunder inside. The rest of the gods headed out as they began to sense the abnormalities happening. The Gods were confused,

but Codeman knew exactly what this was; it was Grim. Grim had arrived.

"Fools you are; you led me right to the portal and did all the dirty work for me. It would have been so much simpler to stash the keys, but you had to be the Hero, didn't you... Mathias... Codeman," rasped Grim. Mathias stepped back, shocked. Even after all the planning, he missed such a simple factor. After seeing Grim, the Gods started heading back, yet again undermining the threat of Grim.

"You're leaving?" asked Codeman, looking back at the Gods.

"Not worth our time," replied one of the Gods as he closed the Gate behind him.

"Guess it's just us, gang," Codeman sighed and announced to his friends. His team nodded, waiting for further instructions. It seemed like Grim hadn't come with any skeletons, so it was a one-man army. However, that was all Grim would need to fight the Champions. As Grim approached the Champions, Codeman explained his plan, *"Neveah, use your powers to jam our brains so that Grim can no longer knock us out. Alon, Zeena, Aria, and I will go for Grim, and the rest of you will protect Neveah. She needs to put all her focus on blocking telepathic attacks from Grim."*

The Champions nodded and got into their positions. Alon, Zeena, Aria, and Codeman charged towards Grim, leaving Neveah and the new recruits behind.

Neveah focused all her powers on defending her team. Thomas immediately ran off deeper into the Realm of Gods. Vanessa used vapors from the sky and turned them into water, made a shield, and turned that into ice. Meanwhile, Aria immediately summoned an army of shadow creatures, some assembled to protect Neveah and the recruits, while some charged toward Grim to overwhelm him. Aria's powers worked best during solar eclipses. After all, during an eclipse, the shadow of the moon would cover every surface, giving her powers plenty of room to breathe; she could also use her powers of hiding in the shadows to basically become invisible and travel from inside them. She used this ability to hide all her teammates. However, Grim could still sense the Aura of the Champions at close range. Zeena used her purified magic against Grim's corrupted magic to create a collision of positive and negative energies; the powers were canceled out; as long as Zeena would output the same amount of energy as Grim, Grim was powerless. Using all these circumstances to their advantage, Alon and Codeman rushed towards Grim for close combat. Alon sweep kicked Grim's staff, causing Grim to lose balance, and Codeman followed that with a punch with his Gauntlet, powered by a strength buff spell. The staff that was in Grim's hands flew out of his hands and landed in Alon's hands. Alon immediately burned the staff to the point; only the crystal remained. Then Alon caught the crystal and stuffed it tightly in his fists, increasing temperature and pressure, causing the crystal to break; the crystal released a ton of corrupted energy, which was absorbed by Alon. The overwhelming amount of corrupted magic that Alon took in had knocked him out. Thomas, the recruit, was watching this from afar. He separated his spirit from his body, arrived on the scene, and took Alon back to where his body was hidden as a safe place for the injured Champions.

Although Alon was off the battlefield, Grim was still at a huge disadvantage, and Mathias had hope that he could beat him. Codeman threw in another punch, and then another, and then another. Grim fell to his feet, backing up slowly, but Codeman kicked him one more time, detaching Grim's skull from the rest of its skeletal body that hid behind its dark purple cloak.

"Did we do it?" Codeman wondered under his breath. *"Did we really win?"* He was amused. The rest of the Champions caught up to him. The Champions felt this relief and excitement; they couldn't believe it themselves; after so long, they had finally beaten Grim, but all their smiles and laughter quickly wiped off their faces. A huge noise was heard from behind the court. The Champions rushed towards the noise they had heard. Behind the court was a huge garden, which had a boundary of hedges and a large gate. The Champions opened the Gate to see many dead Gods and Grim standing over the fountain to fight more Gods. The God of the Cosmos threw moons at Grim, but Grim turned the moons to dust before they could

get near him; the God of Time tried slowing down time to stop Grim's movements and turn back time to when Grim didn't go into the Fountain of Godhood. However, Grim easily brushed his attempts off by speeding up his own movements. With every moment, more and more gods were dropping; as an angel, Grim was already very powerful, but as a god, Grim might as well be unbeatable.

While the Champions were battling it out with Grim, Amy was dealing with the shock of her parents. Memories of her parents kept popping up in her mind, and this particular memory, the scene of when she last saw them, was looping in her head.

'She had just come home from a clinical rotation; on the way home, a huge ship had arrived above Techtopia with an army of Deathbringers lowering down and raiding the city. With all the chaos and confusion, Amy was lost; she made her way back home to see a Deathbringer at the gate of her building. She quickly hid in an alley and looked outside to see the Deathbringer grabbing her parents by the hair, forcing them into their ship, and getting in the driver's seat himself; after that, she hadn't seen her parents since. She was taken in by the Resistance that Aunt Billie formed and then by the Champions as their on-site doctor.'

Amy returned to reality from her flashback, with Leah shaking her and asking if she was okay. *"Where did they find the bodies?"* she asked. Leah nodded and asked Amy to follow her. The two went to a crash site behind a mountain near the graveyard. There used to be a forest in the area, but after the crash of the ship, it had been cleared out. With the impact and the crater, it was obvious the ship was small-scale and used for household travel, similar to an escape pod or an earth car. As Amy got closer to the remains of the ship, maneuvering through the debris that was dug into the ground, she recognized the ship to be her own family vehicle. She wiped the dust off the window and saw the body of a Deathbringer inside, with a screwdriver stabbed into his head. Amy knew that screwdriver was the one her dad always kept on him in case he had any issues with machinery, as he worked in a small repair shop that he owned. Amy could picture the scene of her parents being abducted and taken prisoner by Satan's men, and in an attempt to fight back, they killed the Deathbringer. But with no one piloting the ship, they ended up crashing. Amy wiped the tears streaming down her cheeks. She recollected herself, reminding herself of the bigger task at hand.

"Leah, do you know where I can find a ship? Better if it is able to travel through realms and space," said Amy. *"Well, my uncle owns a rental, so you could get one from there. I can't promise it will be free of charge; he's a bit greedy, but I can try to get you a discount,"* said Leah.

"Thanks… That means a lot. This will help a lot of people, I promise," said Amy. The two headed back to town, where Leah took Amy to her uncle's rental. Amy was surprised to see a huge building with multiple vehicles parked, some of which she had never seen. From hoverboards, hover boots, and hover scooters to jetpacks and jetboots. From huge hovering sports cars and hoverbikes to escape pods and small ships. There were even tanks and helicopters fueled with new laser technology and cars that could transform into boats or planes in a matter of seconds. On top of the building lay perhaps the most advanced ship that Amy had ever seen; the ship was perhaps as big as Mathias's ship and had an equal number of features. The two stepped inside the building, where Amy saw a man screaming into his earpiece, arguing about taxes.

"Uncle Al, I have a favor," said Leah. Al held up his palm and asked Leah to wait. Leah pulled out her digi-wallet, which immediately caught Al's attention.

"Ooo, customers… So what can I do for ya? Make it quick, though, haven't you heard? Time is money?" said Al, throwing his earpiece on the white marble counter.

We are looking for a ship better suited for space travel and maybe on a discount."

"Discount…pft," laughed Al. *"Al's vessel Rental doesn't do discounts…you know that, Leah,"* said Al.

"You don't? Not even if I tell Mom about the spaghetti incident?" said Leah.

"Well, there's no need for that now, is there… Let's see, space travel, you say?" said Al, browsing through keycards on his card display.

"Here, this one should work just fine," said Al, pulling out a keycard and throwing it towards Leah, "Tell you what, you can rent that one for a week for free. Just don't tell your mom about the spaghetti incident, act like it never happened," said Al, rubbing his bald head. Leah agreed, and the two girls walked out.

"What a weird guy," said Amy.

"That's what Uncle Al is known for," said Leah.

"And what's the spaghetti incident?" said Amy.

"Long story," chuckled Leah.

"Wanna talk about it over coffee when I'm done with all this?" said Amy.

"Bet… Just make sure you come back this time," said Leah.

The two girls hugged each other before parting ways. As Amy got ready to leave, Leah warned her about the rules of the rental. "The minute you start the ship, a count will start; it will count down to a week, and as soon as the timer runs out, the ship will autopilot back home. If you wish to extend your duration, the ship will give you a warning before leaving; you must *extend the limit within a minute, or else it will leave. And, of course, you will be charged extra,"* said Leah.

"Thanks," said Amy. And with that, Amy booted up the ship and left for Earth at full speed.

The Champions stayed hidden, hoping the Gods could handle Grim. However, most Gods had failed to do so. As more Gods fell like flies, Codeman, in a blink of an eye, found himself back in the court, this time with the Supreme God standing right in front of him.

"Where am I? Where are my friends?" said Codeman.

"You are in the 5th dimension. Normally, your human mind cannot comprehend this dimension, but I have adjusted your mind to be able to comprehend it and see it in a more familiar form, being 3D," replied God.

"Why did you bring me here?" said Codeman. The God explained,

"Mathias, the realms are in grave danger," the Supreme God began. Mathias rolled his eyes in his mind, thinking, *'Now you understand. As if we were speaking in an alien language a few hours before.'* However, outwardly, Codeman maintained his composure and patiently waited for the Supreme God to continue.

We gods underestimated the power of our own creation, and we need your help. Grim is killing every

last one of us; although immortality means you can't die, it is only limited to aging. Fatal injuries can still lead to death. Nothing was able to do a god that much damage until Grim, which is why I want to make you my avatar. As the Supreme God, I'm not allowed to interfere directly with the physical world or its outcome, but I can assign avatars to do my dirty work; you will be the embodiment and incarnation of me. You will have all my powers and the knowledge to use them; after the task is done, I will withdraw these powers. Do you agree with these terms?" asked God.

"What kind of rule is that? Aren't you the one in charge? Can't you break your own rule? Who's going to impose this rule upon you?" said Codeman.

"My rules impose themselves; when I made this rule, it meant that if it was broken, Godhood would be taken away; why do you think all those Gods out there are losing? Their powers are reduced to magical abilities as they are interfering with Grim's plans directly." Mathias was confused and annoyed at the Gods, but seeing how strong Grim had gotten, he had no other way; he held his hand up and shook it with the Supreme God. As soon as he made contact with the deity, Mathias woke back up in his body. He felt this overwhelming amount of knowledge; it was as if all the secrets of the universe had been revealed to him. He now knew and could unlock every spell and power imaginable.

Codeman walked up to Grim with a new power in him.

"If it isn't Codeman, the pathetic loser who couldn't even tell me apart from my clone," said Grim. Codeman walked closer and punched Grim in the jaw, charged with purified energy, neutralizing Grim's own corrupted magic. Grim reached out his hand and shot beams of corrupted magic, but after being neutralized, he was weaker. Codeman used more purified energy to slowly tear away Grim. Grim was an entity made of pure negativity, although once an angel, after Grim had unlocked free will, Grim lost physical form and became the embodiment of negativity, which made Grim no longer an angel. But negativity can easily be fought back with positivity; that's how purified magic was made. With positive energy, it was easy to fight Grim as long as the output was strong enough to neutralize him. Mathias kept throwing in everything he had, using all the purified magic he could produce, strong enough to tear apart Grim. Grim fell to the floor and started backing up; at this point, this wasn't even a fight or a battle but a beatdown. Grim threw in all the negativity he had through corrupted magic, but Codeman, who was oozing with purified magic, easily deflected it. And with one last kick to his head, Grim was finally dead. Codeman looked up at the sky; the moon continued its orbit, the clouds went back to being fluffy and white, and the sun finally touched the

pearly white walls and the golden gates of heaven. He closed his eyes as he felt the energy leaving his body.

Codeman woke up to see Amy performing a healing spell on him to restore his soul and energy that he had used up in the fight.

 That really drained you; you made it look so easy as well," said Zeena. Mathias looked next to him to see Alon sitting up, producing flames.

"Bro, can you believe it? The corrupted magic I took in purified after you killed that bastard? Now my powers are all enhanced," said Alon.

Mathias smiled, seeing how, after so long, they had finally won. Although Mathias had lost all the skill and energy he had gained while being an avatar, the knowledge had remained, and with enough training, he knew he could reach the same levels of power again, master every spell again, and create his own purified magic.

As the Champions were about to leave, Mathias noticed the fountain was completely drained. Before leaving, the Champions met back with the gods. The gods who had died were reborn with new bodies; however, with the new bodies, they were weaker than before. The Supreme god thanked Mathias for saving the realms but warned him about an enemy. Someone who had stolen the water of the Fountain of Godhood and one who had been after him since the start. "*Satan,*" said Mathias under his breath.

A week had passed now; the citizens of EQ Haven had returned to their homes. Amy had used her magic of healing to fulfill Zoey's request and restore her family from being zombies. Adamina, Amy, Aria, and Mathias studied the skeletons from Grim's army and constructed Robby, a body that could reconstruct itself in different shapes and fix itself if broken. The solar system was finally connected with the rest of the realms with portals built in areas on Mars and Earth.

"*It's been a long few weeks, hasn't it?*" said Mathias. Adamina nodded. The two were walking along the roads of Earth, and Mathias was walking Adamina back to her apartment.

"*You know we could use someone like you on our team. You are pretty gifted with tech, and we could use someone behind all the screens and machinery,*" said Mathias with hope in his voice.

"*It's usually me or Aria with that job, but that distracts us from kicking ass,*" said Mathias.

"*Uh-huh, so you want me to work for you?*" asked Adamina.

"*Well, it's not all work; it would be pleasant to have you around,*" said Mathias, rubbing his fingers together.

"*I'm only kidding; cheer up, Codeman,*" said Adamina, nudging Mathias with her elbow. "*I would *love to join you guys, but it's a long way from home, and there's a lot I'd be leaving behind,"* said Adamina.

"I understand," said Mathias in a somewhat disappointed tone.

"Tell you what, I'll consider your deal and tell you what I think by tomorrow," said Adamina.

"Wanna talk about it over coffee?" asked Mathias.

Sure, pick me up at seven tomorrow. I will choose where. And it's a date," said Adamina. Before Mathias could react, Adamina reached out, gently taking his hand. She traced the lines on his palm with her fingertips, a tender gesture that felt intimate under the starlit sky. With a warm smile that lit up her features, she squeezed his hand, then turned to dash up her apartment stairs, leaving Mathias touched by the simple yet profound connection. With all that had happened and what was going to happen with Satan being missing with the fountain water, Mathias felt himself less worried. Right now, he felt that everything was going to be okay as long as he had his friends with him and Adamina.